P9-DUG-334

SCIENCE'S TRICKIEST QUESTIONS

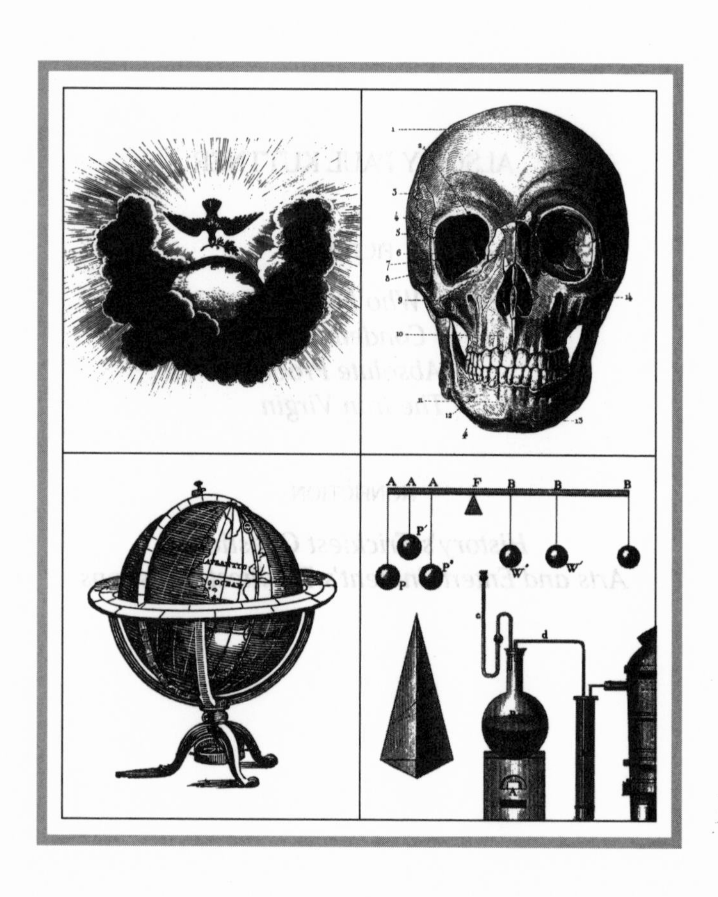
A A A F B B B
P′
P
P°
W°
W′
d

SCIENCE'S TRICKIEST QUESTIONS

402

Questions That Will Stump, Amuse, and Surprise

PAUL KUTTNER

AN OWL BOOK / HENRY HOLT AND COMPANY / NEW YORK

Henry Holt and Company, Inc.
Publishers since 1866
115 West 18th Street
New York, New York 10011

Henry Holt® is a registered
trademark of Henry Holt and Company, Inc.

Published in Canada by Fitzhenry & Whiteside Ltd.,
195 Allstate Parkway, Markham, Ontario L3R 4T8.

Library of Congress Cataloging-in-Publication Data
Kuttner, Paul.
Science's trickiest questions : 402 questions that will stump,
amuse, and surprise / Paul Kuttner.—1st Owl bk. ed.
p. cm.
"An Owl book."
Includes index.
1. Science—Miscellanea. I. Title.
Q173.K98 1994
500—dc20 94-5954
CIP

ISBN 0-8050-2873-0

Henry Holt books are available for special promotions and
premiums. For details contact: Director, Special Markets.

First Owl Book Edition—1994

Designed by Katy Riegel

Printed in the United States of America
All first editions are printed on acid-free paper. ∞

1 3 5 7 9 10 8 6 4 2

To the memory of

Margarete and Paul, my parents, and

Annemarie, my sister,

and also to

Stephen, my son,

and to

Ursula Fraenkel and Ilse Jochimsen.

CONTENTS

QUESTIONS

CONTENTS

ANSWERS

FOREWORD

Science is a discipline that tends to frighten all but its most dedicated students and professionals. Even when one brings the subject down to a more popular level, it can still be demanding, overpowering the reader with too many taxing details. And so I continue my *Trickiest Questions* series and do for science what I have done for history and the arts-and-entertainment world.

Science's Trickiest Questions uses my teasing, tricky style that made its two predecessors, *History's Trickiest Questions* and *Arts and Entertainment's Trickiest Questions,* a success in bookstores, schools, and libraries. This book is filled with 402 scientific queries that will deliberately mislead its audience. Then the answers set things straight, explaining the cause and effect of each query. Also, the answers provide pertinent details that go beyond the question, supplying readers with connected threads and important auxiliary information.

It was my express purpose that this book not deal with trivia such as naming the chemical formula for sulfuric acid, for instance, or identifying two organs of respiration in vertebrates that are situated on each side of their chest. Nor was this book meant to be a tiring crash course treating scientific technicalities. I wanted *Science's Trickiest Questions* simply to be informative, amusing, and fascinating, as it encourages readers to probe further into some of its subjects: astronomy, biology, chemistry, mathematics, physics, botany, and medicine, among others.

But this book is not meant to be a full meal, starting with soup and finishing with dessert. Instead it is supposed to be served as a huge gift box filled with the most delicious bitter and sweet chocolates to be selected arbitrarily. Any hungry mind eager to delve into the secrets of the world can randomly sample this book and learn about the universe and the mysteries within our bodies.

Some questions will be easier than others, as *Science's Trickiest Questions* covers the full gamut of the world's scientific experience, starting with the Big Bang and ending with the incineration of Earth by the Sun. In between, the pages will explore subjects such as climates on other planets, ironies in some scientists' lives, and tragedies involving cancer, AIDS, malnutrition, and overpopulation.

It is by alerting the mind to the scientific mysteries around us that we will learn to better understand the secrets of what makes life meaningful on our planet. Also, by striving to solve and comprehend these riddles of our existence, I hope it will encourage us to preserve life on Earth in all its manifestations.

QUESTIONS

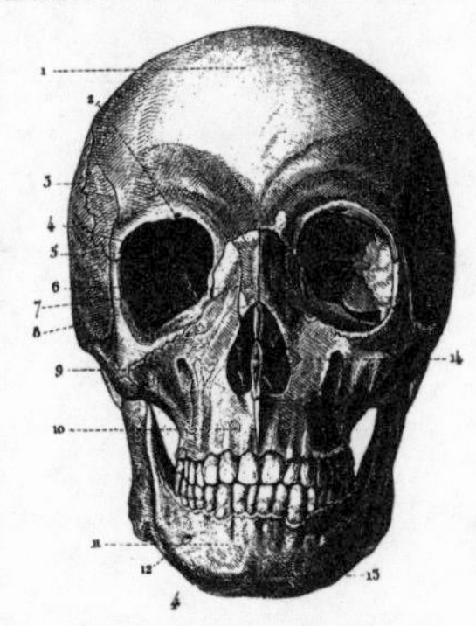

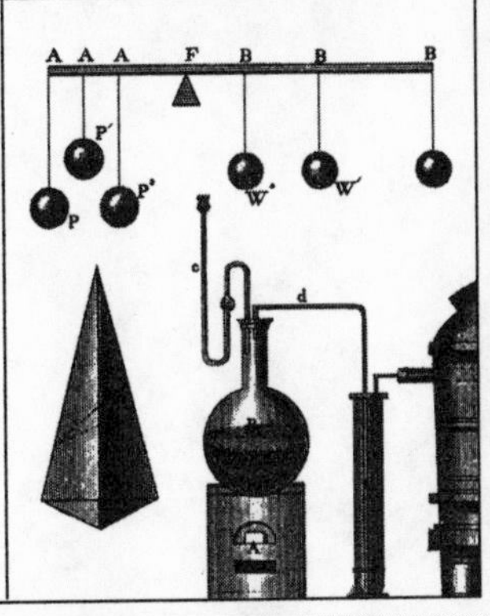

THE NATURAL WORLD

☞ Q 1.

What tree has a caffeine content twice that of coffee? Its fruit is a star-shaped follicle with eight hard seeds, and these seeds, or nuts, contain an essential oil and a glucoside, which are used to make a heart stimulant. Combining its fruit with the extract of another plant, it is used in manufacturing what popular beverage?

☞ Q 2.

Although flies can live through an entire summer, how long are worker ants and their queens known to live? How heavy a load are ants capable of carrying? Five times, ten times, even twenty times their body weight? And South American and African army ants attack what kind of armies?

☞ Q 3.

Can you name a liquid that has been detected in ants and in some plants, such as stinging nettles, and is still sometimes used in the preparation of textiles?

☞ Q 4.

What do Jean Louis Pons (1761–1831) and Carolyn Shoemaker (b. 1930) have in common? Hint: One was a French astronomer, the latter an American amateur astronomer.

☞ Q 5.

You surely know what the following nouns mean, but do you know their meaning when applied to a particular branch of science? Here are the words: hotbed, chat, cap, punk, pitcher, nucleus, mast, and maiden.

☞ Q 6.

Is the science and art of extracting silver from various ores and its reclamation from any type of industrial process referred to as silviculture?

☞ Q 7.

How many hundreds or thousands of years ago did the last giant meteorite crash into our planet? And do we know exactly how many meteorites have hit Earth in the last million years or so?

☞ Q 8.

What genus of the figwort family (*Scrophulariaceae*) that is grown all the way from Central Asia to the Canary Islands is prized as a herbaceous perennial and recognized for the treatment of heart disorders?

☞ Q 9.

Do we know how many meteors have fallen onto this planet from outer space in the last billion years?

☞ Q 10.

It belongs to the family *Psittacidae,* the genus *Melopsittacus,* and the species *undulatus.* It also happens to be one of the most popular pets in the world. What is it? If you have the correct answer, do you know what this pet used to be called?

☞ Q 11.

There are about 5,000 species of this kind. Most of them are reef mollusks, and are collected in every corner of the globe. What are they most commonly called?

☞ Q 12.

When oxygen is bombarded by ultraviolet radiation from the sun anywhere between six and thirty miles (c. 10–50 km) above the Earth, will the Earth's stratospheric ozone layer eventually be depleted to such an extent that the ultraviolet radiation will be a dire threat to life on our planet?

☞ Q 13.

Lampreys are eel-like fish with a suctorial mouth. What do lampreys and sharks have in common virtually alone among all extant fish?

☞ Q 14.

When the Big Bang occurred fifteen to twenty billion years ago, one lump of matter—the kernel of the incredible mass that makes up our universe—exploded, forming today's billions of galaxies. True or false?

☞ Q 15.

What two vegetables differ in the shape of their leaves and have different names, although both have been called coleworts—a name deriving from the Anglo-Saxon term meaning cabbage plants?

☞ Q 16.

Fish and reptiles have something in common with certain dinosaurs as far as their food is concerned. What is it?

☞ Q 17.

Most of us know about the black holes of astronomy: collapsed stars surrounded by a strong gravitational field from which no matter or energy, not even light, can escape. But since we cannot see these black holes, how do astronomers know that they even exist?

☞ Q 18.

What does an ornithologist mean when he says that the lore amounts to an inch and the zygodactyl's circumference amounts to five inches?

☞ Q 19.

Percentage-wise, how much of the power generated by the sun is intercepted by our planet, and what is the name for the power that supplies the sun's energy to the Earth?

☞ Q 20.

How can we determine today what the weather was like a thousand years ago, long before the arrival of newspapers?

☞ Q 21.

Which of the following animals do not develop cataracts during their lifetime: cats, dogs, elephants, horses, or birds?

☞ Q 22.

If somebody told you that nymphs can regenerate their amputated legs, would you consider this part of a weird saga or legend?

☞ Q 23.

What is the difference between an asteroid and a planetoid? And where can they be found?

☞ Q 24.

Here are four figures of Greek mythology: *Europa* was a Phoenician princess loved by Zeus, taking on the form of a white bull; *Io* was also loved by Zeus and changed into a heifer by jealous Hera; *Callisto* (also part of Roman mythology) was a nymph who, because she was loved by Zeus (Jupiter) was changed into a bear by Hera (Juno); and *Ganymede* was a beautiful youth carried off by Zeus to be the cup bearer to the gods. What do these four names have in common in relation to science?

☞ Q 25.

The plants (*Camellia sinensia*) that can grow to a height of thirty feet (9 m) are evergreen trees of the *Theaceae* family. They turn out a full crop of leaves after the first three years of growth about every forty days, and these are picked in Japan and China shortly before they have completed their cycle of growth. In contrast, in Indonesia and India, due to the difference in climates, they are plucked every week or two. What is the end product of this plant's crop in which the processing methods rather than the type of harvested leaf determines its color? Although the plants have been cultivated for about 4,500 years, they do not appear in the Bible.

☞ Q 26.

Can you name the male animal that bears his offspring prior to their birth?

☞ Q 27.

Besides being a first name, what else is Timothy known for?

☞ Q 28.

Can you name the largest star in the constellation *Orion?* Compared to the sun, how big is this star? Is it smaller, or two or even ten times the size of the sun? Is it hotter than the sun?

☞ Q 29.

What is a selenologist? Hint: Among other things, a selenologist studies the terminator and Tycho.

☞ Q 30.

In the botanical department, what is the difference between cacao, cocoa, the cocoa nut, and the coconut?

☞ Q 31.

What edible fruit is native to Africa, consists of fiber, sugar, and water, and has little nutritional value? Some people claim its seeds contain substances that control hypertension, that it has diuretic properties, and that its juice has been used for the treatment of nephritis (disorders of the kidneys). Although the fruit itself is eaten by the millions annually in the United States, its seeds are relished in China and in some Near Eastern countries. In Iran they are toasted, bagged, and sold like popcorn. The fruit's flesh contains 6–12 percent sugar. What is this fruit?

☞ Q 32.

The twelve divisions of the night sky, or if you will, the twelve astrological constellations, are referred to as the signs of the zodiac. Each constellation has its own name. Why do most of their names come from the ancient Greeks? And how valuable are the zodiac denominations to the scientific community of astronomers?

☞ Q 33.

Who first recorded that the number of rings in the cross-section of a tree trunk tells its age? Hint: It's the same person who detected that the moisture of the year's season could be gleaned from the cross-section of a tree trunk by measuring what?

☞ Q 34.

What is the largest order of insects worldwide? And how many known species are there of this order: 90,000? 120,000? 150,000? More? Or less?

☞ Q 35.

Do we know what happened ten-billionths of a second after the Big Bang? Or even earlier?

☞ Q 36.

The order *Diptera* includes more than 85,000 species. Can you name four that are best known among them?

☞ Q 37.

Why is the outer layer of the sun—the corona—much hotter than the outer part of the main body of the sun, the photosphere? And how long would it take for you to be burned to a heap of ashes if you stayed in the corona, which can reach temperatures of several million degrees?

☞ Q 38.

He was a Swede but is best known by his Latin name. Although he wrote the monumental works *Genera Plantarum* and *Systema Naturae,* he is even more famous for what other scientific system?

☞ Q 39.

A butterfly as a rule emerges from its chrysalis in about two weeks (or not until the following spring). Horseflies usually develop from egg to adult in about two weeks as well. How long does it take for the periodical cicada of North America to complete its metamorphosis?

☞ Q 40.

In what way does the orbit of Pluto differ from that of any of the other planets in the solar system? And is Pluto three or four times the size of Earth, or a third or quarter of Earth's size?

☞ Q 41.

We all know that peaches spoil more quickly than other fruits sold in the market because they are not waxed as are say, apples, pears, and cherries. Is there another reason for their faster spoilage?

☞ Q 42.

Last century, some birds were known by the following names: a) golden-winged woodpeckers, b) butcher-birds, c) rice buntings, d) great-footed hawks, e) Carolina pigeons, and f) cock-of-the-plains. What are they known as today?

☞ Q 43.

When the sun and the planets of the solar system were created out of the gaseous state of interstellar dust between four and five billion years ago, what percentage of this mass ended up in the sun and what percentage in the planets?

☞ Q 44.

What are the two most puzzling things about the planets Venus and Jupiter?

☞ Q 45.

Dr. William L. MacDonald, of West Virginia University in Morgantown, and one of the world's leading experts in his field, forecast in 1992 an event that turned out to be the greatest ecological tragedy in modern times. What was it? Hint: By exposing the affected areas to a less virulent strain of Cryptogamia he claims that still greater ecological damage can be contained and averted.

☞ Q 46.

What is the common but inaccurate name for a well-known bird—*Haliaetus leucosephalus*—and where can the bird be found?

☞ Q 47.

Are fish cold-blooded?

☞ Q 48.

Can you name the planet that is about the same size and density as the Earth? Its orbit is closest to our planet, yet its surface temperature is about 900°F (468°C) and its atmosphere consists of about 95 percent carbon dioxide (Earth's is about 78 percent nitrogen). It is the slowest rotating body in the solar system (one day there is the equivalent of 243 Earth days), so slow that it actually spins slowly backward. Astronomers refer to it as an *inferior planet.* Why? And what is its name? Hint: When we get up early in the day, we may spot this planet and call it by another name.

☞ Q 49.

Apart from a lunar eclipse, has there ever been a time when the moon was totally invisible to the naked eye on a cloudless night?

☞ Q 50.

Lately, many Hawaiian and South American restaurants list dolphin on their menus. Why doesn't anybody protest when man's favorite friend, *Flipper,* is killed for human consumption?

☞ Q 51.

What is a dendrochronologist? And what is perhaps the hardest thing he or she has to decipher?

☞ Q 52.

Is it science fiction if you read that Swift-Tuttle may pay us a visit in the years 2126 and 3044? Or that we'll run out of carbon dioxide and life on earth will cease to exist in a hundred million years?

☞ Q 53.

Is the red part of the strawberry a fruit or a vegetable?

☞ Q 54.

Have brown dwarfs been integrated in their community? Or has it been scientifically established that the brown pigment is a warmer shade of white?

☞ Q 55.

If somebody told you that a "fixer" is necessary to keep you alive, would you agree?

☞ Q 56.

Touch any snake and you will note that it is cold-blooded. So how do snakes warm their eggs?

☞ Q 57.

Which of the following two scientific theories was proven correct 1.) the geocentric or 2.) the heliocentric?

☞ Q 58.

The world's largest flower (or bloom, to be more precise) is named after whom? And why can it also be considered one of the world's most unpopular plants?

☞ Q 59.

What animal other than a human is known to shed tears for emotional reasons?

☞ Q 60.

Since 1885, the Little Magellanic Cloud galaxy (LMC), is the first supernova that could be seen with the naked eye and the brightest since 1604 in our terrestrial skies. Keeping in mind that light travels 186,282 miles per second, how many miles is the LMC galaxy from the planet Earth?

☞ Q 61.

The hummingbird is distinctive in two ways. What are they?

☞ Q 62.

We have all been bitten by mosquitoes, but what is the primary purpose of the enzyme in a mosquito's bite that causes the resulting itch?

☞ Q 63.

Can we still see a supernova that was first noted by the Chinese, the Japanese, and the Native Americans in the Southwest almost a thousand years ago, in 1054?

☞ Q 64.

Name the mammal, weighing anything from ¾ to 2½ ounces, that consumes almost its entire weight in food every day. Except for the consumption of insects, this mammal is a vegetarian. It builds and maintains its own pathways, about the width of a garden hose, to connect its "sleeping accommodations" with its feeding grounds. Hint: The mammal is so cute that one of the world's most famous motion picture "stars" is based on its likeness.

☞ Q 65.

What unique phenomena occurred in 1805 and 1935, which are not expected to happen again until the year 2160?

☞ Q 66.

A member of the genus *Pulex* and the genus *Culex* are most annoying to the human race. Add to them the genus *Pediculus,* and you'll have three dreadul creatures, whose depicted life cycles and illustrations three and a half centuries ago were so phenomenal that they have never been equaled. Who was responsible for the masterful delineations of these abominable creatures, and what are they?

☞ Q 67.

How many light years away from Earth is the Andromeda galaxy? Is that more or less than the distance between the Little Magellanic Cloud galaxy and Earth?

☞ Q 68.

Can you name the plant that is known both for its narcotic and its poisonous properties and also has been used to dull pain during surgical operations? In addition it has been used as an aphrodisiac and was supposed to have a unique habit that Shakespeare even considered worthy of mentioning in *Romeo and Juliet.* How exactly did he describe this incredible habit?

☞ Q 69.

All of us have seen cows chewing their cud, but do they actually ruminate while the grass passes through their stomach?

☞ Q 70.

There is one plant in the botanical kingdom that bears a name based on an expression of surprise. In its lifetime, this plant changes from white to pink, ending up blood-red. Can you name it?

☞ Q 71.

Can you name the star or stars that supply our planet with one-fifth the amount of light that the full moon sheds on Earth?

☞ Q 72.

What has a mass of just over 6.5 sextillion (6.5 followed by 36 zeroes) U.S. short tons and moves at the rate of 1,100 miles per minute?

☞ Q 73.

The planet Mercury, even its poles, is bathed in boiling-hot sunlight of up to 800°F. (c. 430°C). The planet, after all, is only one-third as far from the sun as is Earth. Yet NASA's Jet Propulsion Lab claims in 1992 that there is ice on the planet. How can that be explained?

☞ Q 74.

What contains calcium, sulfur, magnesium, some trace elements, carbon dioxide, and phosphorus?

☞ Q 75.

From about twenty-five to seventy-five of these go off each year, and they release more than 10^{38} joules of electromagnetic radiation. Do we refer here to underground or atmospheric nuclear tests? A joule (J) represents the energy expended in one second by a current of one ampere at a potential of one volt.

☞ Q 76.

Can you name the tree that in the eighteenth and nineteenth centuries was an excellent source of timber? Its fruit was harvested to be turned into flour and feed livestock, and its wood and bark were used in the production of huge amounts of tannin, a chemical used to tan leather.

☞ Q 77.

It is the most concentrated form of any known organic food and contains eighteen of the known twenty-two amino acids, including all the eight essentials, making it a complete protein. In fact, it produces twenty times more protein than do soybeans grown on an equal land area. It contains nutrients unlike any other single grain, herb, or plant, such as gamma-linolenic acid, arachidic acids, beta carotene, high amounts of B_{12} and iron, RNA and DNA nucleic acids, and chlorophyll. It also aids in protecting the immune system, in mineral absorption, and in cholesterol reduction. What is this miracle drug?

☞ Q 78.

There are three main entities around a particular star. Two of them are called the photosphere and the chromosphere. What is the name of the third? And exactly where are all three located?

☞ Q 79.

What is unique about the grunion?

☞ Q 80.

What do these three elements have in common: isotopes of helium and hydrogen and a tiny amount of lithium-7, which consists of a nucleus that has four neutrons and three protons?

☞ Q 81.

The Greeks and Romans gave this lovely body a name, meaning wearing "long hair." What is it? Hint: It can be extremely long, with a heavenly shape.

☞ Q 82.

How much larger is the sun than our planet in mass and diameter? What are the sun's two most prominent gaseous elements? And can we predict what will happen to that star after it doubles its present age?

☞ Q 83.

Although the following men made their names in various scientific fields, what else did they have in common: Norbert Rilleux, Granville T. Woods, Henry Blair, William A. Hinton, Ernest E. Just, Charles Drew, Daniel Hale Williams, Percy Julian?

☞ Q 84.

True or false? Our planet has a steady magnetic field. If this is not true, does the Earth's magnetic field reverse, and why?

☞ Q 85.

Centipedes belong to the class of arthropods, a phylum of invertebrate animals with jointed limbs and segmented bodies. How many legs do they have?

☞ Q 86.

What kind of energy, in conjunction with water and carbon dioxide, is the driving force behind the growth of plants?

☞ Q 87.

What happened before the Big Bang—let's say 25 billion years ago—and who first referred to this cataclysmic event as the Big Bang?

☞ Q 88.

What are the two most prominent characteristics of butterfly fish?

☞ Q 89.

How did the German astronomer and mathematician Johannes Kepler (1571–1630) make more perfect sense of the Copernican theories of the heliocentric universe? And who based some of his own theories on Kepler's data?

☞ Q 90.

Nature as we know it wouldn't exist without xylem and phloem. Exactly where do you find these two entities, and what functions do they perform?

☞ Q 91.

When you see sunspots on photos or in films, is their dark color enough proof that these sunspots are either black or dark brown?

☞ Q 92.

The scientific establishment believes that this particular virus is a member of the rhinovirus subgroup of picornaviruses. Yet even though the medical profession has no clear understanding of its precise source, a serious outbreak of this virus's contagious disease in almost each case requires the slaughter of the infected animals, who can transmit this dreaded disease to humans. What is this disease best known as?

☞ Q 93.

During World War I, wounded soldiers were frequently left unattended on the battlefield, and their open sores were invaded by immature flies or maggots. These infested wounds often cleared up faster than those of other soldiers whose wounds were insect-free. What was the reason for this strange occurrence?

☞ Q 94.

In the constellation Perseus you may be able to locate the winking star Algol, which is Medusa's left eye in the night sky. If you watch this star for a few days in the autumn sky you will note that Medusa is actually winking at you. Has anybody ever been able to figure out why Algol is a "winking star"?

☞ Q 95.

What branch of science is particularly interested in the period/luminosity relationship of Cepheids?

☞ Q 96.

Although most sports fans know that Jack Dempsey was the world heavyweight champion in boxing from 1919 until 1926, when he lost the title to Gene Tunney, is it true that Jack Dempsey is also related to a scientific field?

☞ Q 97.

How poisonous is malic acid, which is a crystalline hydroxy dicarboxylic acid, $C_4H_6O_5$?

☞ Q 98.

How much of a smile can you expect to get from a smilodon?

☞ Q 99.

Suppose you put a carrot with its leaves attached in a plastic bag inside the crisper (bottom drawer) of your refrigerator and then store another carrot without its leaves in a plastic bag next to the first. Punch a few holes in the plastic bags and keep them in the refrigerator for about a week. Will there be a difference in the taste of the two carrots at the end of the week? And if so, why?

☞ Q 100.

Is Alexander's Dark Band a forerunner of Irving Berlin's "Alexander's Ragtime Band"?

☞ Q 101.

Carolus Linnaeus completed his development of the use of binary nomenclature in botany *Species plantarum* (*Species of plants*) in 1753. It is still the foundation for the classification of botanical species. Had anyone ever tried to introduce a classification scheme for plants before Linnaeus, or even succeeded in the attempt?

☞ Q 102.

We have seen the Southern Cross and Centaurus, especially its brightest star, Alpha Centauri, on the firmaments of many a planetarium. But have they ever been seen with a naked eye outside a planetarium?

☞ Q 103.

One of the great naturalists in the last century and a half was an abolitionist and an active participant in the Underground Railroad, sheltering fleeing slaves in his family's house. More than 130 years after his death a newly discovered work of his about the rise of new generations, nature, and science was published by a small press in Washington. Who was this naturalist and philosopher? Hint: He was the first Anglo-American field ecologist to be influenced by Darwin's theory of natural selection.

☞ Q 104.

Although this perennial herb's essential oils are used in chewing gum, liqueurs, candies, and even medicines, the plant is much more famous because its leaves produce something known the world over for its soothing effect. Hint: In Greek mythology the god of the underworld, Pluto, fell in love with a nymph whose name gave the abovementioned herb the last syllable of *its* name. What is the name of this plant?

☞ Q 105.

Why are our planet's seasons relatively stable, with the Northern Hemisphere experiencing warm weather as it leans toward the sun in the summer months, while six months later it is the Southern Hemisphere's turn to lean toward the sun?

☞ Q 106.

Can you name one of the most popular plants of the Western world, but one that botanists refer to as parasitic? Even though this plant has its own chlorophyll, it grafts or annexes itself by means of an *haustoria* (Latin for "sucking mouth") to the branches of trees, sponging on the chlorophyll there. In the western United States, it is often found on oak trees, while on the Eastern seaboard, it is usually attached to maples, linden, poplars, and apple trees.

☞ Q 107.

Which of the following two words about foliage is spelled correctly: obovate leaf or ovate leaf? And what does the correct word mean?

☞ Q 108.

Believe it or not, one of the world's most popular flowers was named about 2,000 years ago by the Greek medical writer Dioscorides for its bulbs' resemblance to a man's testicles. What is this beautiful plant called in the English-speaking world?

☞ Q 109.

Are there any openings in an egg's shell before you yourself break it open or before it is accidentally cracked?

☞ Q 110.

The first and second souchong and the first and second congou are part of what well-known plant?

☞ Q 111.

Can you define a common and then a more complex term: jacket and rhizobium? And do you know in which scientific discipline they are used? In chemistry? Astronomy? Or in another scientific field?

☞ Q 112.

What great scientist consented to the publication of his masterpiece only when he was already on his deathbed?

☞ Q 113.

It is an established fact that in summer microbes release carbon dioxide by breaking down organic matter in the soil. We know that photosynthesis shuts down after sunset. Is it also true that in dark, frozen winter conditions microbial activity grinds to a halt in the dirt?

☞ Q 114.

What relatively underrated and now little-known astronomer paved the way for the work of Johannes Kepler (1571–1630) and provided irrefutable evidence for the new heliocentric system, enabling Kepler to formulate his three laws of planetary motion—a superlative achievement that realistically described the heliocentric system proposed by Poland's Copernicus (1473–1543)?

☞ Q 115.

The scientific Latin name is *Salix babylonica.* Are any of these still around or did they die with the demise of ancient Babylonia thousands of years ago?

☞ Q 116.

Almost three thousand years before Christ the star nearest the north pole was Alpha Draconis (Thuban). Will it still be nearest to us in, say, twelve thousand years? If not, what is the reason for the change?

☞ Q 117.

Is the passionflower, a plant of the genus *Passiflora,* named so because eating these herbaceous vines or herbs arouses a sexual appetite?

☞ Q 118.

Do you know the names of the two best-known epiphytes? And even though you may be able to find them anywhere, they are natives of which part of the world?

☞ Q 119.

Do cicadas belong to the same order as locusts?

☞ Q 120.

With all the waste gas of carbon dioxide (CO_2) we exhale at every breath (about 4 percent of our exhalation), there must be a surfeit of CO_2 in the air around us. Percentage-wise, how much of the exhaled carbon dioxide do we breathe in again?

☞ Q 121.

Do the names of the garden plant with pendulous, funnel-shaped flowers, the fuchsia, and the magenta dye of the rosaniline series, fuchsine, derive from the Latin or the Greek? On the average, how tall does the fuchsia usually grow?

☞ Q 122.

What do bottlenose dolphins and bats have in common?

☞ Q 123.

Can you name the sizable animal whose relatively primitive immune system prevents it from getting sick or infected as a result of inflicted wounds? What is the reason for this miracle? And can any part of this animal be applied to the treatment of cancer in humans since these huge creatures are rarely known to be afflicted by cancer?

☞ Q 124.

When people in the scientific field talk of "naked smut," would they look for an answer to this sorry plight in the works of Jung, Adler, or Freud?

☞ Q 125.

How long can plant seeds remain viable? Will they still sprout after ten years? After a hundred years? After five hundred years? Even longer? Or does their ability to grow in a particular climate cease well before any of the above allotments?

☞ Q 126.

Is it an old wives' tale that birds' stomachs will explode if you feed them rice? Should you only feed them birdseed?

☞ Q 127.

What is the huge matter scientists refer to as *macho* when they are not even sure that it may exist?

☞ Q 128.

How many products did one American scientist make from the *Leguminosae* family? Who was that scientist? Hint: *Leguminosae* is an order of herbs, senna, shrubs, and trees bearing legumes.

THE HUMAN BODY

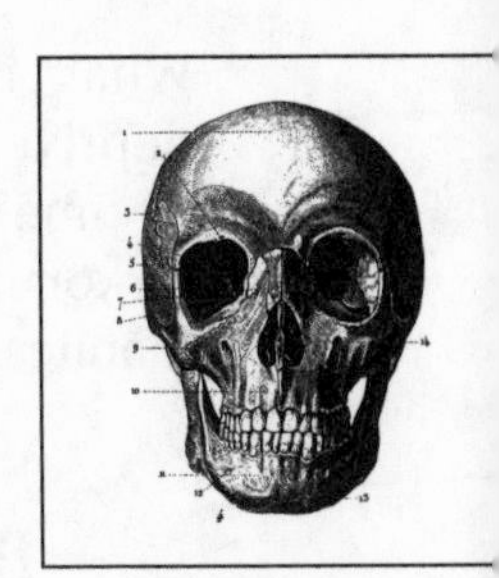

☞ Q 1.

Is the primary purpose of vitamins to make the human body stronger?

☞ Q 2.

This simple medical procedure has been done billions of times without complication on males, but performed on the opposite sex it can have dire, often fatal, consequences. It is considered cruel, senseless, and illegal in many countries when performed. What is it?

☞ Q 3.

It's a long word, but you couldn't very well exist without what it stands for. What exactly is the function of cholechromopoiesis?

☞ Q 4.

What is the result when humans are completely deprived of niacin and protein over a prolonged period of time? Is the resultant debilitation known as kwashiorkor in some parts of the world? If not, what are the manifestations of kwashiorkor?

☞ Q 5.

When speaking of loops, composites, whorls, and arches in one branch of science, what exactly are we referring to?

☞ Q 6.

Why are amalgams so useful in dental work?

☞ Q 7.

Why is mixoscopia related to watching movies that have raised eyebrows since the beginning of the twentieth century?

☞ Q 8.

Is there any well-known vitamin that the human body cannot produce on its own by chemical reaction, although other mammals can still do so independently?

☞ Q 9.

What do you call the nucleus of human red blood cells? Exactly where are these cells created?

☞ Q 10.

Can you name the virus that is a leading cause in the United States of liver cancer, liver failure, and cirrhosis, yet can remain in the blood of an infected patient for up to twenty years without causing any signs or symptoms of liver inflammation?

☞ Q 11.

During sexual intercourse, is it safe to prevent the transmission of HIV-positive cells by using a condom just prior to male ejaculation?

☞ Q 12.

Since nicotine is known to be harmful to the human body, how much more pernicious is nicotinic acid, and what organs does it attack?

☞ Q 13.

Almost every living thing has them. The highest number can be found in *Ophioglossum reticulatum* (a species of fern) and the smallest amount in *Myrmecia pilosula* (a species of Australian ant). What are they? And why does every living thing have twice as many haploids as diploids, even the abovementioned fern and ant?

☞ Q 14.

Why are fruit flies ideal for the study of genetics?

☞ Q 15.

What is the difference between the thymus gland and the thyroid gland?

☞ Q 16.

With all the new medication on the market, has skin cancer (melanoma) decreased perceptibly in the last sixty years?

☞ Q 17.

On what occasion do cells manufacture proteins that are known to invade adjacent cells in humans? And what is the reason for this strange diffusion of proteins?

☞ Q 18.

Has it ever been proven why thousands of people suffer from headaches when they go skiing?

☞ Q 19.

By nature, the botanical bearers of sperm have one goal: to reach the stigma, the moist, sticky receptor of the female organ of a flower. But for a very definite reason millions of people curse at least one of these bearers. Why? A hint: Medical doctors refer to it as allergic rhinitis, and 22 million Americans suffer from it.

☞ Q 20.

Micrographia, the most important work by English microscopist Robert Hooke (1635–1703), was published in London when he was only thirty years old. When Hooke had glanced through the microscope at a piece of cork he noticed something there to which he gave a Latin name that has since become a very common English word, especially in scientific circles. What was the word?

☞ Q 21.

After conception in a female body, what operation during cell division ceases temporarily?

☞ Q 22.

The postulates (fundamental assumptions) of what scientist have been accepted for more than a hundred years as the criteria necessary to prove the etiology of diseases caused by fungi, viruses, bacteria, and protozoa? And how did this scientist go about proving his theory?

☞ Q 23.

Approximately how many cells are there in the human body?

☞ Q 24.

Why does "sand" gather in the corners of our eyes when we are tired? Why does the feeling of sleepiness produce this sand?

☞ Q 25.

When we hear of free radicals, we may think of members of a left-wing political group roaming about freely in society. But what is the meaning of the term "free radical" in scientific fields?

☞ Q 26.

How do skin pores in your scalp and hair follicles differ? Or do they differ at all?

☞ Q 27.

What did Sigmund Freud consider to be his greatest work?

☞ Q 28.

Even though hydrochloric acid kills, in what way can it be useful to the well-being of the human body? Hint:

Enzymes battle your food right after hydrochloric acid has done its "dirty" work.

☞ Q 29.

Why can a person eat certain foods for the first twenty or even thirty years of his or her life and suddenly become allergic to them?

☞ Q 30.

Can you name something that is unique in your life, yet measures only about a millionth of an inch across? At the same time, if you'd be able to stretch it to its full length, it would extend more than three feet.

☞ Q 31.

This is not a miracle drug, but a puzzling miracle chemical, which is produced by adrenal cells located above the kidneys and transmitted from there to the brain. As little as we understand about its properties, we know that we cannot function without it. Precisely what does this chemical, dopamine, do for us? And what do scientists hope its chemical substitute can perform that dopamine can't?

☞ Q 32.

What exactly is the relatively new medical procedure called phacoemulsification and how did it lead to the even more up-to-date procedure referred to as tunnel incision?

☞ Q 33.

You have virtually none of these in your bladder and lungs, yet hundreds of millions are active in the alimentary tract, many millions at the base of your eyelashes, and they love to take a shower with you. Dr. Theodor Rosebury has stated that they were responsible mostly for stimulating man's immunity mechanism. What was he referring to?

☞ Q 34.

In all fairness, who should be called the father of psychoanalysis? And what does this question have to do with Freud's termination of friendship involving three colleagues?

☞ Q 35.

In the field of medicine, what does SIDA stand for?

☞ Q 36.

What do the following long-winded terms have in common: positron emission tomography (PET), magnetic resonance imaging (MRI), single photon emission computerized tomography (SPECT), and superconducting quantum interference device (SQUID)?

☞ Q 37.

Why does taking a few aspirin sometimes delay or inhibit the cure of colds or flu, even though swallowing them makes us feel better?

☞ Q 38.

Are any vitamins manufactured within the body?

☞ Q 39.

Where do you find the hippocampus (major and minor)? In Africa? In the world of botany? Inside your body? Or in the night sky?

☞ Q 40.

Cataracts must be swollen and opaque before an ophthalmologist can remove them. True or false?

☞ Q 41.

Past the age of fifty most men will have an enlarged prostate, the principal storage depot for seminal fluid. By the age of seventy, the prostate will have enlarged by fifty percent. What causes this enlargement?

☞ Q 42.

What breakthrough studies credited to Sigmund Freud came outside the fields of psychology and psychoanalysis?

☞ Q 43.

Only 2 to 5 percent of the human race has one simian crease, while the rest of us have two. Exactly what is a simian crease? Hint: It's located below the dermal ridges.

☞ Q 44.

Cholesterol is a constituent of every human cell membrane and it remains in a latent, inactive state when mixed with cells in a laboratory. Why does it elicit blockages and lesions on the linings of artery walls? Can anything be done to prevent complete blockage of the arteries?

☞ Q 45.

Is there any internal part of the body that cannot develop cancer?

☞ Q 46.

Can lasers be used to remove cataracts?

☞ Q 47.

Why do bones "crack," or make an explosive sound, when people crack their knuckles?

☞ Q 48.

Virtually every teenager and adult has been in touch with a sphygmomanometer. What is its function?

☞ Q 49.

What is salty, bitter, sour, and sweet?

☞ Q 50.

The disease pellagra is caused by a dietary deficiency characterized by what physical disturbances?

☞ Q 51.

How many light receptors are located on the human retina, and what color will be most difficult to see as a person ages?

☞ Q 52.

What medications are produced from willow bark?

☞ Q 53.

What is the most common sexually transmitted disease: AIDS, syphilis, or gonorrhea?

☞ Q 54.

How many sperm are released into a woman's vagina during ejaculation by the male, and how many of them reach the egg (ovum) to attempt fertilization?

☞ Q 55.

In the 1980s, what happened when young Asian immigrants in the United States experienced trouble expressing themselves verbally in English in college?

☞ Q 56.

What part of the human body can increase up to 200 times its normal volume?

☞ Q 57.

Who can be given credit for eradicating yellow fever in Cuba (where it had been endemic for centuries) and for creating public works and improving sanitation and education?

☞ Q 58.

Does semen flow at the same speed all the way to the woman's egg?

☞ Q 59.

Why was Sigmund Freud afraid that he would die in 1918, the last year of World War I?

☞ Q 60.

What are the primary functions of human blood vessels, the arteries, veins, and capillaries? How do they differ?

☞ Q 61.

What antibiotic has been proven the best remedy in fighting against the common cold?

☞ Q 62.

Nazis always accused the Jews of dominating the medical profession in Germany before Hitler came to power in 1933. What percentage of medical doctors were Jewish before 1933 in Germany?

☞ Q 63.

What hormone's principle function is to regulate the condition of the inner lining (endometrium) of the uterus? Who discovered this hormone?

☞ Q 64.

After strenuous exercise, what accumulates in the muscles?

☞ Q 65.

Why can diabetes affect a person's vision?

☞ Q 66.

What are the only areas of the human exterior that do not sweat?

☞ Q 67.

Which of the following parasites are known to continue living within the human body after a person has died: tapeworms, maggots, hookworms, and/or roundworms?

☞ Q 68.

Which of the following aberrations of the eye are the first accompanying symptoms in the incipient stage of glaucoma: flashes of light, distorted color vision, fuzzy vision, or floating spots?

☞ Q 69.

Can one define a genome not only as a maxim, an aphorism, and a misshapen dwarf that supposedly inhabits the interior of the earth, but also as a noun with a scientific connotation?

☞ Q 70.

What exactly is the method "polymerase chain reaction" that is currently used by police labs to hunt down suspected criminals?

☞ Q 71.

It has been medically established that diabetes that first appears in adulthood is closely related to weight problems. How much of this ailment is due to faulty insulin production by the pancreas?

☞ Q 72.

It is a well-established fact that when we have a fever or an infection, iron levels drop significantly in the blood. What is the reason for this sudden drop? Where does the iron disappear to? And how many iron supplements do we need to replenish the supply while we are sick?

☞ Q 73.

How important is the substance called adenosine triphosphate, or ATP, to the maintenance of our physical health?

☞ Q 74.

What is a capnometer? Hint: The person benefiting from it isn't even aware that he is being helped.

☞ Q 75.

What gland in the human body contains a fluid that makes the vaginal canal less acidic and helps to transport, support, and nourish the fragile male sperm?

☞ Q 76.

Most spiders are venomous, while insects like bees, ants, and wasps are much less poisonous. Has it ever been determined how many more people die from spider bites than from insect bites?

☞ Q 77.

About nine months after conception, just prior to childbirth, the mother's anterior pituitary gland produces prolactin and other glands produce a watery liquid called colostrum. This fluid contains antibodies and protein that help to protect the newborn from infection. Three days later, the colostrum is replaced by what?

☞ Q 78.

Is it a biological fact or an old wives' tale that people with a lower iron content in their blood suffer fewer heart attacks than do those with a higher iron content?

☞ Q 79.

What exactly are histones, and what functions do they perform?

☞ Q 80.

Their entire lives are spent copulating. Two hundred million people around the world can tell you about it from first-hand experience, and it was first discussed in writing at the time of the Egyptian pharaohs about four thousand years ago. What exactly is it?

☞ Q 81.

What is the hardest substance in your body, and what is its main component (96 percent)? Your skull? Your toenails? Or something else?

☞ Q 82.

When a spine is injured severely, destroying nerve cells for about forty-eight hours after an accident, does the body's immunological system spring into action and attack renegade oxygen molecules that are eating through the spinal cell membranes?

☞ Q 83.

What is the common ancestor of viruses, and in what way are they phylogenetically related?

☞ Q 84.

Which award above any other did Sigmund Freud (1856–1939) value most in his life? Can you name the scientific findings for this award?

☞ Q 85.

When a person is declared dead, it can be assumed that right after somatic death, the patient's brain and heart have ceased to function. Is the above statement true or are there any human cells that continue to live following somatic death?

☞ Q 86.

What did Louis Pasteur (1822–1895) discover when he inoculated old cells of cholera bacteria into chickens?

☞ Q 87.

In what physical condition do human antibodies refuse to fight off a disease and instead merge with the affected genetic material?

☞ Q 88.

Can any other parts of the human body be affected by gum disease and badly decayed teeth?

☞ Q 89.

Is it an old wives' tale that the consequences of eating too much food—even if it is good, healthy food—can affect your teeth adversely, regardless of whether you brush them several times a day?

☞ Q 90.

Why do our hands turn blue when we shiver with cold? And why do our faces turn red when we perspire?

☞ Q 91.

Can blood coming from a number of healthy donors, all of who have a different blood group, ever be safely injected into a human patient with a blood group different from that of the donors?

☞ Q 92.

Could you live without splanchnology?

☞ Q 93.

What stimulating drug decreases appetite but causes an increase in motor activities and in the ability to concentrate? It is also able to contract blood vessels in the mucous membranes and to relax smooth muscles in the lung, turning it into an effective nasal decongestant. It tends to produce a heightening in blood pressure and a change in behavioral patterns. Since it also releases dopamine, a neurotransmitter, in the cerebral cortex and norepinephrine from nerve terminals it can, with the latter, increase blood pressure by forcing the muscles of the heart to contract more forcefully. What is this drug, and has it ever been abused by those who take it? If so, why?

☞ Q 94.

If somebody tells you that you have coccidioidomycosis, would you consider this a reason to celebrate?

☞ Q 95.

What do you call the blood test used to detect gonorrhea? And are there already nonculture tests that detect gonococcal antigens directly in specimen material when diagnosing male and female genital infections?

☞ Q 96.

What was the greatest killer disease in the last two hundred years and, consequently, can it be called the worst of all malignant ailments?

☞ Q 97.

In what way are diabetes and gum disease connected?

☞ Q 98.

Most drugs are synthesized in factories today. But how does the effectiveness differ in drugs originally developed from botanical sources (e.g., penicillin) and those developed from chemical sources (e.g., sulfa drugs)?

☞ Q 99.

When they are eaten alone, the calcium in spinach and chard is not as easily absorbed as is the calcium in milk. What can be done about this?

☞ Q 100.

What would have happened to you if your parents had altogether lacked the following neurochemical cousins of amphetamines: norepinephrine, dopamine, and, foremost, phenylethylamine (PEA), with a good touch of oxytocin, followed by a latecomer called endorphin?

☞ Q 101.

What do the Hungarian obstetrician who pioneered in asepsis (freedom from pathogenic microorganisms), Ignaz Philipp Semmelweis, and the United States' Oliver Wendell Holmes have in common?

☞ Q 102.

How many red blood cells are normally destroyed by leukemia per day in a patient afflicted with this blood cancer?

☞ Q 103.

Ptyalin has made life sweeter and more pleasant for almost everybody. What is this thing called ptyalin? Is it a chemical instigator of love? An ingredient of ice cream? Part of the fragrance given off by blooming plants? Or something else?

☞ Q 104.

If HIV is the virus that causes AIDS, where does it hide out and remain dormant for up to ten years before it breaks out and becomes deadly? Can drugs such as AZT benefit those with HIV?

☞ Q 105.

About 99 percent of an ingredient that is also used in the manufacture of enamel, glass, cleaning agents, lime, and marble belongs to that part of the human body that keeps us in motion and enables us to enjoy food. Where in the body is this 99 percent located and of what exactly is it made?

☞ Q 106.

For decades it has taken up to eighteen weeks to determine the exact drugs to use in the treatment of a given strain of tuberculosis. What did scientists discover in 1993 that can cut this waiting period in half? Hint: The new scientific discovery, without exaggeration, sheds some light on the patient's medical problem and its solution.

☞ Q 107.

Do people mostly lose their second set of teeth because the roots become slack and loose with age?

☞ Q 108.

What common medical aid was discovered in 1993 to be the culprit in fighting the beneficial effects of penicillin and other antibiotics such as streptomycin and tetracycline when these antibiotics were meant to destroy bacteria that causes, for example, urinary tract infections?

☞ Q 109.

Is there a special human organ that destroys red blood cells, and if so, how many new ones can be replaced per minute in what other part of the human body? What is the main function of red blood cells?

☞ Q 110.

What ailment calls for the most frequently performed surgical operation in the 1990s: heart disease, cancer, or the urinary problems affecting elderly Americans?

☞ Q 111.

When the American scientist Benjamin Minge Duggar and some of his coworkers checked soil samples for antibacterial action in 1944, what did they discover?

☞ Q 112.

What is the difference between a thrombosis and an embolus? And what do we know about the man or woman who coined these two medical terms?

☞ Q 113.

What Englishman was the father of a medical innovation that changed the history of medicine late in the eighteenth century? He struggled for decades to have his invention (related to cattle) accepted, although it was rejected by the Royal Society (because of faulty discoveries by other more established doctors). Early in the nineteenth century, finally, a parliamentary grant was voted for the maligned doctor and his theory was accepted as a medical practice that would come to be used throughout the world. Who was this doctor and what did he discover?

☞ Q 114.

Before there were even one billion people on Earth, what great economist suggested that "moral restraint" might prevent famine due to overpopulation? Exactly how has the world fared in eliminating famine since then?

☞ Q 115.

What was the greatest irony about Adolf Hitler (1889–1945) being saved from total blindness after he and his company were subjected to an Allied chlorine gas attack near Ypres a month before the November 1918 armistice?

THE HOME PLANET

☞ Q 1.

Has the mercury in a thermometer ever hit below –70°F in a permanently inhabited place in Siberia?

☞ Q 2.

If you measure a mountain or volcano from its base to its summit, what is the highest mountain on Earth?

☞ Q 3.

An enormous coral reef almost encircles this island, which extends about five miles (eight kilometers) from north to south and three miles (five kilometers) from west to east. Captain James Cook first described it in

1769. Home to picturesque bays, coral reefs, blue lagoons, and coconut palms, it is about 135 miles (220 kilometers) northwest of Tahiti and is so incredibly beautiful that it has attracted thousands of tourists and artists annually and served as a location for motion pictures. Can you name the island and what it was millions of years ago?

☞ Q 4.

Can you name the most important changes our oceans will undergo due to global warming?

☞ Q 5.

We have learned about the Paleozoic Era, the Precambrian and Mesozoic eras. Can you pinpoint the approximate age of the Cenozoic Era? In what geologic period were most of our existing mountains formed?

☞ Q 6.

Can you name the year(s) when the United States experienced the hurricanes Stella, Theodore, Ursula, and Xavier?

☞ Q 7.

What is the difference between a tropical cyclone, a typhoon, and a hurricane? When was the last time, for example, that there was a hurricane in *The Wizard of Oz* state of Kansas or a monsoon in or near California?

☞ Q 8.

Where and what is the highest navigable lake in the world? How and why do the inhabitants of its region differ physically from those living at lower altitudes? Hint: It is the legendary birthplace of the Incan civilization.

☞ Q 9.

What happens when underground pressure and intense heat is exerted upon fine-grained limestone over a period of millions of years?

☞ Q 10.

Today we know that about 70 million years ago, at the end of the Cretaceous period, the age of the dinosaurs ended abruptly after they had dominated the planet for about 70 million years. Has it ever been scientifically established that another comet or asteroid obliterated most of life on this planet before or since the demise of the dinosaurs?

☞ Q 11.

Name the one thing the *ornithomimus* and the *garudimimus* had in common, and to a lesser degree the *oviraptor.* And why is the most famous of their kind frequently misnamed?

☞ Q 12.

Not every animal that has ever lived has been fossilized for us throughout the ages. Most of them have decomposed since they died without being buried. According to today's paleontologists, how many animals or species do you think have perished in the last 248 million years or so, of which there is no known fossil record? Do you think that more than one animal species in a dozen, or in a hundred, or in a thousand has perished without leaving a trace?

☞ Q 13.

All of us have seen drumlins. But what exactly are they?

☞ Q 14.

What toxic solution shaped the subterranean chambers at the near million-year-old Carlsbad Cavern that was carved out of solid rock and lies about 1,000 feet below the ground of New Mexico?

☞ Q 15.

In 1988 the first fiber-optic cable across the Atlantic could carry approximately 40,000 simultaneous conversations. A 1990 optical-fiber cable could double this achievement. It is not possible to eavesdrop on anything passing through these cables, making their use ideal for transmitting national-security messages. When was the first transatlantic cable laid to transmit telegraph messages between what geographical points?

☞ Q 16.

Who first charted the Gulf Stream? And by how many degrees warmer can it be than the surrounding waters of the Atlantic Ocean?

☞ Q 17.

If somebody told you that about 15 percent of Earth's terrain is made up of karst, would you dispute this fact or show off and maintain that the hills of Slovenia do not make up fifteen percent of the Earth's surface?

☞ Q 18.

Who were the famous participants in the oldest case about surrogate motherhood, a case in which an illegitimate boy's birth raised enmity between him and his father's rightful son? Hint: This conflict still hasn't been resolved.

☞ Q 19.

Did the Three Mile Limit law in any way prevent a recurrence of the kind of nuclear near-disaster that happened at the Three Mile Island nuclear plant?

☞ Q 20.

Creationists insist that Eve was the first woman and the mother of the human race. Do twentieth-century biologists agree?

☞ Q 21.

After the *Titanic* struck an iceberg on April 14, 1912, taking 1,522 of the 2,227 passengers aboard to their deaths the following morning, how many other ships are known to have struck a berg, resulting in fatalities? A dozen? More? Or fewer?

☞ Q 22.

Of all the mass extinctions on our planet in the last billion years, how many can be credited to impacts by asteroids or comets, to huge catastrophic lava eruptions, or to a combination of both?

☞ Q 23.

On June 18, 1858, Charles Darwin (1809–1882) received an article from the naturalist Alfred Russel Wallace (1823–1913), who was then ill in Ternate in the Molucca Islands. The article summarized Wallace's theory of evolution by natural selection. How can it be proven that Darwin did not plagiarize some of Wallace's theories in his own *On the Origin of Species*, which was not published until November 24, 1859?

☞ Q 24.

Has the cause of the formation of life on Earth ever been determined?

☞ Q 25.

His book *Guide to Geography* is full of factual errors. Many of his calculations of longitude and latitude are faulty, particularly involving the measuring of distances, because he based his theories on Poseidonius's estimates, which were 30 percent short in accuracy. This astronomist places the equator too far north, and there are representations at variance between the text and the maps in the *Guide,* where he omits mention of climate, inhabitants, and products while his analytical observations regarding mountain ranges and rivers are flawed. Yet he is considered one of the great and most influential scientists of all time. Who was he, and what is known about his life?

☞ Q 26.

If part of the Earth continues to move 5 centimeters (2 inches) a year for the next sixty million years and ends up in Alaska, where can this turtlelike motion be detected today? And why?

☞ Q 27.

The solid organic matter of sedimentary rock, frequently called kerogen, can for the most part not be dissolved in petroleum solvents. What happens to kerogen when it is heated?

☞ Q 28.

Were the two terms *survival of the fittest* and *evolution* (in the biological sense) coined by Charles Darwin (1809–1882) or by Alfred Russel Wallace (1823–1913)?

☞ Q 29.

How did the giraffe acquire its long neck?

☞ Q 30.

When he was discovered in 1891, he was erroneously considered the "missing link" between man and ape in the popular press. What do anthropologists call him this century, and how long did he wander this earth? Is it known whether modern man evolved from this "missing link"?

☞ Q 31.

Why do traces of iridium play such a pivotal role in the extinction of the dinosaurs?

☞ Q 32.

What happened at the end of the Cretaceous period about 65 million years ago that has puzzled scientists for generations? Hint: The exact location where something drastic happened was on the northern edge of the Yucatán Peninsula in Mexico.

☞ Q 33.

How much of the Earth's continental surface is made up of sedimentary rock, how much is igneous, and how much metamorphic? Can you give an example of each of these three kinds of rock?

☞ Q 34.

What superlatives do the administrative capital city of Bolivia, La Paz, and the autonomous region of Southwest China, Tibet, have in common?

☞ Q 35.

When scientists speak of the "Cambrian explosion," are they referring to a nuclear device or an event more closely allied scientifically to the Big Bang explosion?

☞ Q 36.

Suppose you spend a winter in Siberia and the thermometer hits forty below zero—not uncommon in that part of the globe. Do you consider it more likely that the temperature is correct if the thermometer reads it as Fahrenheit or as Celsius? Please note that water freezes when the Celsius thermometer hits zero degrees while the Fahrenheit thermometer registers the freezing of water when it is thirty-two degrees above zero.

☞ Q 37.

Early last century it was used as a medicine, but in 1859, after being extracted from the earth, it was refined by some Pennsylvania businessmen who used it to see the world a bit clearer. The same product helped us to move faster toward the close of the nineteenth century and it became as coveted as gold and diamonds. What is this treasure over which wars have been fought for over a century? Hint: Today, millions tend to attack the deleterious consequences of using this material, even though it may have helped to save the free world.

☞ Q 38.

If a man were kept in a doorless, windowless room that contained a bathtub, a bed and a table, how would he know whether he was imprisoned in the Northern or Southern Hemisphere?

☞ Q 39.

What does a site in southeastern Ontario, Canada, called Sudbury, have in common with a site near Iowa City, Iowa? Hint: Both are hundreds of millions of years old.

☞ Q 40.

Is there a place on the equator where both penguins and fur seals can be found?

☞ Q 41.

Is their some archeological mystery about the collapse around 2200 B.C. of the Akkadian empire—the world's earliest noteworthy civilization? The Akkadians were a northern Semitic people who conquered the Sumerians around 2350 B.C. and ruled Mesopotamia. The imperial city of Akkad was founded by Sargon I and became the world's oldest imperial center in the third millennium B.C.

☞ Q 42.

Does the motion of the planet Earth differ from artificial motion?

☞ Q 43.

England's Arthur C. Clarke (b. 1917) is recognized as one of the greatest science-fiction writers of the twentieth century. What farfetched idea did he propose at the end of World War II that actually became an indispensable fact of life throughout the world twenty years later?

☞ Q 44.

As the magma rises and bubbles up, the crust is forced upward and the inner pressure drops, causing emissions within to escape. Sounds almost like baking a cake or a pie, but does the above describe a pleasant activity?

☞ Q 45.

It has long been the conviction of scientists that termites evolved from the ever-changeless cockroaches. Is this belief still scientifically valid?

NUMBERS AND FORMULAS

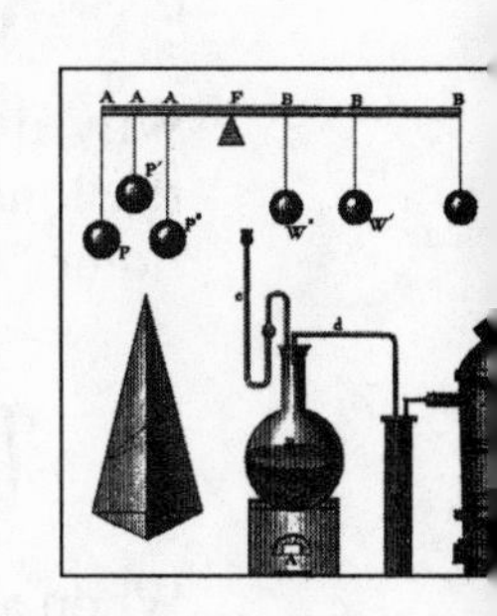

☞ Q 1.

For years we have read about the tragic depletion of the ozone layer in the stratosphere endangering life on Earth. Do we know whether this savior of life, ozone, can be transported to our planet to help us survive?

☞ Q 2.

In navigation the position of a vessel generally is determined by advancing the craft's previous position to a new one. What happens when the captain must determine his ship's position by its speed and on the distance it has traveled after leaving its last port of call? It must also be kept in mind that the captain has to consider the force of the wind, the water's currents and possible compass errors. What is the nautical term for finding such a ship's location, and how was this navigational phrase arrived at?

☞ Q 3.

Why do billions of people throughout the world owe the long-forgotten German physicist Joseph von Fraunhofer (1787–1826) untold gratitude?

☞ Q 4.

When a waiter in a fine restaurant opens a bottle of wine and lets you taste it to see whether the wine is to your liking, what would an experienced oenophile consider wrong with this procedure?

☞ Q 5.

What alloy of tin and copper was so important and left such a permanent imprint on civilization thousands of years ago that an entire era was named after it, even though the time of that era is calculated differently in different parts of the world?

☞ Q 6.

When was the moving picture invented?

☞ Q 7.

Can you guess who made the following predictions?

A) This person claimed there would be no reason for any individual to have a computer in his home.

B) This scientist insisted that expecting a source of usable power from the transformation of the atom was talking moonshine and that the energy produced by breaking down the atom was a very poor "kind of thing."

C) This inventor said in a *New York Times* article on February 25, 1957, that man would never reach the moon regardless of any future scientific advances.

☞ Q 8.

There is one amazingly simple number that, if multiplied by 9, 18, 27, 36, and multiples of 9 through 81, will always result in a total in which each numeral is exactly the same as the one preceding and following it. What is the number multiplied by 9, 18, 27, 36, etc., and what are the totals in which the congregate sums of the numerals amount to the number to be multiplied?

☞ Q 9.

What world-famous scientist made the following statement in 1945?

"I never thought that we would make an atomic bomb, and at the bottom of my heart I was really glad."

Was this statement dissimulation and hypocrisy or an honest admission of incompetence?

☞ Q 10.

What do you get when you chill and compress liquid carbon dioxide below –109° Fahrenheit?

☞ Q 11.

A Scottish botanist observed something in 1827 while watching microscopic pollen grains subjected to a continuous random motion in a fluid medium. He noticed that the hotter the water the faster the grains moved and that as the water cooled the grains slowed down. This odd observation was explained seventy-eight years later in a theory that fully described how the grains were driven by water molecules and that these molecules moved faster as the water heated up. Who first discovered this motion, and who explained in the twentieth century that molecules, of which all matter is composed, are continually in motion? Hint: The botanist was the first to describe and name the cell *nucleus* and he was a pioneer of plant classification, and the twentieth-century scientist was during his lifetime a citizen of three countries.

☞ Q 12.

What is the heaviest known vector meson with a mass greater than ten times that of the proton? Its elementary particle is identical to the nucleus of the hydrogen atom, and along with neutrons, it is a constituent of all other atomic nuclei.

☞ Q 13.

Do you know who coined the word *quark* for the hypothetical particle believed to be the fundamental units of other subatomic particles?

☞ Q 14.

If a young American exchange student, John, told his English host in the British capital that his father had already lived a billion seconds, would his London host be entitled to call his young American guest a fool and a liar?

☞ Q 15.

What do you call the unit of power that is equal to 33,000 foot-pounds per minute in the United States, and why is the correct answer referred to as such?

☞ Q 16.

When was the first color photograph taken? In World War I? In the 1920s? Or earlier? Is the pioneer who took it a little-known scientist, or is his work still taught at universities?

☞ Q 17.

From an architectural point of view, is there a difference between a church steeple and a church spire?

☞ Q 18.

Is the opposite of bottomology topology? And what do the two terms signify?

☞ Q 19.

When we light a candle, is it the wick or the wax that provides the flame?

☞ Q 20.

How long does a championship boxing match last that goes for the full length of fifteen rounds? Figure that, as usual, each round lasts three minutes and the rest periods between the fifteen rounds last one minute each.

☞ Q 21.

When somebody discusses 162 blocks of sarsen, what is he referring to?

☞ Q 22.

Can you name the inventors and scientific discoverers of the following achievements?

1) He discovered the first law of elasticity for solid bodies.

2) Hundreds of years ago he outlined for the first time in an almost precise manner the theory of universal gravitation.

3) He discoursed on the application of barometrical indications to meteorological forecasting.

4) He invented the wheel barometer.

5) He originated the idea of using the pendulum as a measure of gravity.

6) He is responsible for the application of spiral springs to the balances of watches.

☞ Q 23.

As recent as in the 1980s, nitric oxide was considered a toxic molecule. Would you say that even in tiny amounts NO, a colorless gas, should be considered detrimental to the nervous system of *Homo sapiens?* Or is it more apt to damage the bodily functions of humans?

☞ Q 24.

Is light a particle or a wave?

☞ Q 25.

What happens when crystals of silver halide (a binary salt formed by the combination of silver and any of the negative elements fluorine, chlorine, iodine, bromine and astatine) are exposed to light?

☞ Q 26.

What is made up of leptons, hadrons, mu and tau mesons? And what is the purpose of the two latter entities?

☞ Q 27.

Who made the following statements in 1945 about the building of the atom bomb?

1) "I am thankful we didn't succeed."

2) "We didn't do it because all the physicists didn't want to do it on principle."

☞ Q 28.

Why does it help us to know that all living things absorb carbon from carbon dioxide in the atmosphere?

☞ Q 29.

What chemical element has provided us with the following three common uses?

A) It helps to keep us warm.

B) It is a property that we share with our automobiles.

C) It is something people have even killed for.

☞ Q 30.

An engineer in Nazi Germany and another in Great Britain worked independently of each other on a technology that, if the German had succeeded, could have tipped the military balance of World War II in Nazi Germany's favor. In the 1930s they had succeeded in their primary scientific attempts, but their inventions were not fully realized during the war. Today, a world without their inventions is unthinkable. Who are these two men, who later became close friends and now reside and work in the United States? The ex-Nazi is now a U.S. citizen. What did they invent? Hint: Their invention saves everybody a lot of time.

☞ Q 31.

Frederick Scott-Archer (1813–1857) and Richard Leach Maddox (1816–1902) were largely responsible for what has become the most widespread hobby of the twentieth century. Exactly what did these two Englishmen invent?

☞ Q 32.

This scientist discovered the fundamental law that bears his name, and he also articulated a scientific principle that carries his name in reverse. Who was he and what did he discover?

☞ Q 33.

When a doctor uses light amplification by the stimulated emission of radiation, will the patient run the danger of suffering from radiation sickness?

☞ Q 34.

Who expressed these sentiments in a love letter to his mistress at the beginning of the twentieth century? "How happy and proud I will be when the two of us together will have brought our work on the relative motion to a victorious conclusion!"

☞ Q 35.

Its chemical formula is $CHCl_3$, and it was first made in 1831 simultaneously by Samuel Gutherie in the United States, E. Soubeiran in France, and Justus von Liebig in Germany, but Sir James Simpson of Edinburgh was the first person to practice with it on a patient in 1847. What is this chemical formula commonly known as? Hint: The word is made up of the Greek word for pale-green or grass-green and of the Latin word for ant.

☞ Q 36.

Exactly how far does light travel through a vacuum in 1/299,792,458 of a second? And why is this precision timing so important?

☞ Q 37.

What is the name of the eighteenth-century metaphysical thesis that claims everything consists of innumerable, indivisible units with self-governed individual properties that determine their past, present and future? And in what language was it written?

☞ Q 38.

How much heat is needed to melt dry ice?

☞ Q 39.

In the third century B.C. Apollonius of Perga introduced three mathematical-geometrical concepts in his book

Conics. By cutting a cone in certain ways, he arrived at some geometrical shapes. He referred, in Greek, of course, to the slope of the cuts as something that either "exceeded" or "equaled" or "fell short" of the slope of the cone. What are these three words in quotes known by in English from a geometrical point of view?

☞ Q 40.

Was molten lead as a means of fusing pipes in the plumbing trade first introduced to the general public in the eighteenth, the seventeenth, or the sixteenth century? Or much more recently? Or even earlier?

☞ Q 41.

Alan Turing, the scientist who helped to break Nazi Germany's "Enigma" code during World War II, is considered by many to be the father of the twentieth-century digital computer (1936). Why is he associated with the apple in *Snow White and the Seven Dwarfs?*

☞ Q 42.

By how many degrees does the average desk fan cool the air when turned on full speed?

☞ Q 43.

What is ten brix in July, fifteen brix in August, and twenty-two brix in September?

☞ Q 44.

The 1986 round-the-world flight on a single load of fuel by Dick Rutan and Jeana Yeager on *Voyager* was how much faster per hour than Charles Lindbergh's trans-Atlantic flight of the *Spirit of St. Louis* in 1927?

☞ Q 45.

When Albert Einstein lay dying in Princeton Hospital on April 18, 1955, what were the last words he uttered to the nurse assigned to him?

☞ Q 46.

Although Jacques-Yves Cousteau is best-known for his underwater documentaries, he should also be remembered for what scientific invention in 1943?

☞ Q 47.

In 1901 Ernest Rutherford (1871–1937) and Frederick Soddy (1877–1956) discovered that thorium, left unattended, actually changed into another element. What did these two great scientists fail to realize at the time? And what word did Soddy introduce to the English language in connection with the above?

☞ Q 48.

Sir Isaac Newton deduced the laws of gravity and optics, and invented calculus. Yet his masterpiece,

Principia, which reigned supreme for about three centuries after its third book was printed in 1687, is known to contain some mathematical errors. True or false?

☞ Q 49.

Name the countries that have not formally begun converting to the metric system. When was it first proposed by a United States government official that the United States adopt a decimal-based system?

☞ Q 50.

At fifteen, a boy who hated school was left behind by his parents when they moved from Munich to Italy because his father had financial trouble with his factory. After following his parents and enjoying himself for a year in Italy, the boy went back to school, this time in Switzerland, where he became a Swiss citizen in 1900. He married a Serbian girl, Mileva Marić, in 1903 and they had two sons. When he could not get a satisfactory job, he wrote four papers and his dissertation for his doctorate. Who was he? Have any of these papers been preserved for posterity?

☞ Q 51.

In 1792, Antoine Laurent Lavoisier, the father of modern chemistry, who explained the exact nature of air and coined the term *oxygen,* was accused of what heinous crime by a famous figure of the French Revolution?

☞ Q 52.

The word hexagon is derived from the Greek *hes,* meaning six, and *gonia,* meaning angle—a six-angled figure. Pentagon is a five-angled figure, the Greek word for five being *pente.* What are the names of two geometrical figures that derive from Latin equivalents?

☞ Q 53.

What is the strand of continuity running through the four items below?

1) Pierre Curie's hands were swollen by exposure to radium.

2) A pilot in World War II named a B-29 bomber after his mother. What was her name?

3) A Nobel Prize winner from Italy overheard the desk officer at the U.S. Navy Department in Washington in the late 1930s tell his admiral, "There's a wop outside." Who was the Italian?

4) At the end of the nineteenth century, a scientist working in Montreal, Quebec, discovered that uranium had a penetrating radiation that could pass through a sheet of aluminum several millimeters thick. A year later, with Frederick Soddy, he proposed the modern theory of radioactivity. Who was this man?

☞ Q 54.

The American founder of the National Academy of Design worked on what single-circuit electromagnetic contraption in 1832, succeeded with it in 1836, and sent the words, "What has God wrought!" on his controversial invention in 1844? Hint: The machine still carries his name today, although the U.S. Supreme Court had to step in and clear his name for the development of his other, even more famous scientific invention.

☞ Q 55.

Most materials expand as a result of higher temperatures. Is this true of water when it is heated?

☞ Q 56.

Scientists everywhere consider this 600-page scientific book perhaps the most influential ever written. Only a few hundred copies are sold every year, although it has never been out of print in over 300 years. What is it?

☞ Q 57.

What attracts lightning to the leaves and branches of a tree?

☞ Q 58.

By studying and writing extensively about the flight of birds, was Leonardo da Vinci (1452–1519) on the right track in developing a flying machine? Did he ever try to design a helicopter?

☞ Q 59.

Thomas Savery and Thomas Newcomen devised a steam engine in 1705 for what practical purpose?

☞ Q 60.

What scientist contributed to the first successful uranium and thorium fission in Nazi Germany in 1938 with Nobel Prize–winner Otto Hahn (1879–1968) and Fritz Strassmann (1902–1980)? Hint: This scientist coined the term "splitting the atom" in 1939.

☞ Q 61.

Who wrote the following: "Let no man who is not a mathematician read the elements of my work"? Planck? Einstein? Plato? Steinmetz? Leibniz? Newton? Faraday? Archimedes? Or somebody else?

☞ Q 62.

What element is present in virtually all organic compounds?

☞ Q 63.

What did Ernest Rutherford deduce in 1909 when one of his physicists, Ernest Marsden, determined that some alpha particles bounced back from a thin gold foil?

☞ Q 64.

What is the ideal height for a television satellite to be placed above the equator to rotate at the same speed as Earth and be able to relay pictures that have been beamed into it back to another place: 998 miles, 4,000 miles, 11,000 miles, 22,300 miles, or 91,100 miles?

☞ Q 65.

Lavoisier (1743–1794) wrote his thesis on oxygen (although based on an erroneous supposition that the formation of acids depended on a union with a nonmetallic body) in 1775 and presented it to the Academy of Sciences in 1777. Did anyone before him realize that a relatively small amount of air (oxygen) kept blood red and supported life? How was this scientifically proven?

☞ Q 66.

Has it ever been established when and how the measurement of feet (as in footage) was devised?

☞ Q 67.

A few years after immigrating to the United States from Germany in 1889, a young scientist co-authored *The Theory and Calculation of Alternating Current Phenomena* in 1897. He wrote many textbooks that are still used in colleges throughout the world. He became a teacher, General Electric's consulting engineer, was registered with 195 patents, and although a cripple and never married, he adopted a boy at the age of forty. Who was he? What inventions is he best known for?

☞ Q 68.

When English-speaking scientists talk about quattuordecillion, what do they refer to? And why could the end result of their discussion about the above be confusing?

☞ Q 69.

Why is the saline content of the Dead Sea so much higher than that of the Atlantic Ocean? How much salt can be found in a ton of water in the Dead Sea?

☞ Q 70.

How can it be determined through Johann Wolfgang von Goethe's writings that he was less successful with physical than with physiological optics, and why would physicists find it impossible to implement his argument today?

☞ Q 71.

Since the photoelectric effect could not be explained adequately by the laws of classical physics, Albert Einstein made use of another scientist's theory to describe it. Later he adapted that scientist's theory to his own vision of the structure of the universe. Then, in 1913, Niels Bohr incorporated this same scientist's theory into his evolutionary analysis of atomic structure. After Bohr's work was completed the scientist was awarded a Nobel in physics (1918). Who was he, and what theory is he most acclaimed for today? Hint: The scientist's son, Erwin, was executed by the Nazis in 1945 for taking part in a 1944 plot on Hitler's life.

☞ Q 72.

When a scientist maintains that 273.15 is the equivalent of zero, what does it mean?

☞ Q 73.

In 1888, a German physicist incidentally discovered something without realizing its potential. Two coils of fine, but broken wire were lying side by side in his laboratory. When the scientist discharged into one coil the electric power from a Leyden jar, it caused a spark to fly in the second coil too. Who was the scientist and what significance would his discovery eventually have?

☞ Q 74.

One of the most common household items in use since the 1960s was accidentally invented in the 1930s. It is made of fluorocarbon resins that are derived from petroleum and coarsened in a process similar to sandblasting with aluminum oxide for the most part, after which a primer is applied and coated. Where in our homes do we use this item, and what is it?

☞ Q 75.

How can you explain that a boy born in the nineteenth century would be eight years old on his first birthday and twelve years old on his second birthday?

☞ Q 76.

Almost all of us have physical contact most of our waking hours with perchloroethylene, or C_2Cl_4, a compound that is 14.48 percent carbon and 85.52 percent chlorine. Is this chemical formula part of what we eat or what we breathe?

☞ Q 77.

J. Robert Oppenheimer (1904–67) and his Los Alamos physicists knew in 1945 that a new force would be unleashed on the world with the atom bomb. They also knew it would require extraordinary skill to keep it in check, and that the citizenry should be educated about the dangers and opportunities presented by nuclear energy. For technical and humanitarian reasons he felt that the hydrogen bomb should not be built. Likewise, about 400 years earlier, a scientist refrained from having his treatise published because he felt it would be against his Christian principles to use his theory in the killing of Christians. Who was the scientist, and what did he not want to see published, only to change his mind five years later?

☞ Q 78.

What happened, over a century and a half ago, when copper plates were coated with silver iodide and developed by exposure to mercury vapor?

☞ Q 79.

What happens when the bottom of clouds accumulate a large excess of electrons?

☞ Q 80.

Is it possible for electrons to leap from one orbit to another, in small amounts, without passing through the space between them?

☞ Q 81.

He won a Nobel Prize for a theory he claimed nobody really understands. Instead of using the customary calculations, he invented his own nuclear-particle squiggles and arrows that first puzzled everyone but then proved so workable that physicists now accept them as regular scientific entities. What did he win the Nobel for, who was he, and how did he teach classes to his doctoral students? Hint: He determined the cause of the *Challenger* space disaster that killed its entire crew on January 28, 1986.

☞ Q 82.

When turning on a light switch, the electrons in the current produced inside the generator of the electrical plant will flow to the light bulb at an incredible speed of just over 186,000 miles per second. True or false?

☞ Q 83.

How can you explain the fact that when you buy a suit or a dress it seems to be a different color in the store than outside on the street?

☞ Q 84.

Sometime between sixteen and twenty billion (milliard in the U.K. and Germany) years ago the universe was created, as the Big Bang theory makes clear. Even today we can still hear a sort of an echo from that cataclysmic event. What exactly is the nature of that echo, and is it one of the unsolvable puzzles of cosmology today?

☞ Q 85.

When professionals compare the 18 million pixels of their trade to the usual 20 million silver molecules seen in their craft, and the better-quality 100 million silver molecules enjoyed in the dark, what exactly are they referring to?

☞ Q 86.

Who first published instructions on how a computer could be made to work?

☞ Q 87.

Somebody once wrote that if you jump up and down in the cabin of a moving ship you will hit the floor at the same place you would land if the ship were stationary. You will not make larger jumps toward the stern nor toward the prow despite the fact that during the time you are in the air the floor under you may be going in a direction opposite to your jump. To somebody observing this jump from the shore it will appear that the jumper is not landing in the same spot but moving in the direction of the ship's motion. To an onshore observer it will also appear that the deck underneath a jumper on a docked ship will move in a direction opposite to the jump. In both cases the descriptions of the jumper are relative to the situation of the observer. Can you name the person who clarified this paradox in his thesis?

☞ Q 88.

In what field of science do you find beauty that is virtually invisible, and where no beauty exists it's referred to as a hidden state?

☞ Q 89.

What is Einstein's Theory of Invariants?

☞ Q 90.

What invisible, odorless gas is considered a major contributor to the greenhouse effect, accounting for almost 25 percent of global warming? Hints: Reducing this flammable gas by about 20 percent would help to cut levels of global warming. It not only derives from forest fires, coal mines, burning rice paddies, and landfills but also from the digested waste of the approximately 3½ billion animals such as cattle, goats, camels, and sheep, as well as from the decomposition of vegetable matter.

☞ Q 91.

Who made the following two statements at the beginning of the twentieth century (1908)?

1) “Nobody has ever noticed a place except at a time, or a time except at a place.”

2) “The views of space and time which I wish to lay before you have sprung from the soil of experimental physics, and therein lies their strength. They are radical. Henceforth space by itself, and time by itself, are doomed to fade away into mere shadows, and only a kind of union of the two will preserve an independent reality.”

☞ Q 92.

What is mostly composed chemically of compounds of hydrogen and carbon and often contains sulfur, either uncombined or present as a part of certain hydrogen-carbon-sulfur compounds? Hint: In its pristine state it is composed of thousands of different chemicals, including gases that are largely dissolved because of the extreme pressure to which this chemical is subjected.

☞ Q 93.

An amalgam is an alloy of mercury and any other metal. Only one common metal will not form an amalgam or alloy with mercury. What is this exception?

☞ Q 94.

What do the Riemann Hypothesis, Goldbach's Conjecture, the Poincaré Conjecture, and Kepler's sphere-packing problem have in common?

☞ Q 95.

What is the difference between –273.15° Celsius and –459.69° Fahrenheit, and why are these temperatures important for scientists?

☞ Q 96.

What is the difference between baking a cake 1,000 feet above sea level and baking one at an altitude that is below 500 feet?

☞ Q 97.

How can you figure out the following multiplication in your head without much trouble?

$$74 \times 86$$

And what would be a relatively similar computation that you could guess without resort to a multiplication table?

☞ Q 98.

What world-famous scientist maintained as late as 1945 that an atom bomb would require tons of uranium 235 rather than a few pounds?

☞ Q 99.

How can you put nine barrels of oil in two even rows of five barrels each? And how can you line up eleven barrels of oil in two even rows of seven barrels each?

☞ Q 100.

What do the beginning steps toward the unification of algebra and geometry as well as the solution of the general cubic equation of the third degree have to do with "a jug of wine, a loaf of bread, and Thou"?

☞ Q 101.

Leonardo da Vinci and the Chinese scientist Wan Hu independently designed something in 1500. Their two inventions have much in common and are indispensable to the life of the twentieth century. How successful were their designs and what was the end product of their inventions as they are used today?

☞ Q 102.

James Watt is the first person to use steam heat by installing steam pipes in his office. The shrapnel shell is invented by Henry Shrapnell, and so is the threshing machine by Andrew Meikle. A working model of a steam-powered carriage is built by Scotland's William Murdock. The first ascent with a hydrogen-filled balloon is achieved by England's Vincent Lunardi. Bifocal eyeglasses are ushered in by Benjamin Franklin. The puddling method, stirring up molten iron in a furnace to convert it into wrought iron, is developed by the Englishman Henry Cort at Fontley Forge, Hampshire. (Cort is also responsible for developing to perfection the rolling mill with rollers that are grooved.) What do all these inventions have in common? Hint: Henry Cavendish announces for the first time what water is composed of, and in Germany the poet Johann Wolfgang von Goethe discovers the intermaxillary bone, which is located in the upper jaw of humans but not present in most other mammals.

☞ Q 103.

Has it ever been possible to prove the validity of Einstein's General Theory of Relativity (1915) in a laboratory test?

☞ Q 104.

Scientists have often considered stoichiometry undefinable, yet it can be explained in many different ways because it applies to so many different facets of what branch of science?

☞ Q 105.

Did anybody honor or credit the French author Jules Verne (1828–1905) posthumously for anything he wrote?

☞ Q 106.

A wine merchant has two large bottles only, one holding six gallons, the other ten gallons. The glass of the bottles is so dark that it is impossible to see their contents. Moreover, the outside of the bottles is not calibrated, so it is also impossible to discover how much liquid is inside each bottle. A customer, however, insists on acquiring only eight gallons of wine. How does the wine merchant fill the ten-gallon bottle with exactly eight gallons?

☞ Q 107.

Where did Thomas Alva Edison (1847–1931) go to college and earn his degree in science, particularly in physics? Or did he get his B.Sc. for his studies in chemistry?

☞ Q 108.

When the wind blows from the south, does the rooster's head on a weathervane point north or south? And why have roosters (of all the animals in the animal kingdom) frequently been chosen to grace the top of church spires as weathervanes?

☞ Q 109.

The German mathematician and astronomer Johannes Kepler (1571–1630) discovered it in 1609, but unbeknownst to him the scholastic philosopher, scientist, and theologian Albertus Magnus (c. 1200–1280) wrote about it in 1260, and even before him some Chinese scholars noticed it in 135 B.C. What these great scientists, as well as the French mathematician and philosopher René Descartes (1596–1650), noticed was that one of nature's most beautiful sights almost universally shared one thing, even though the thousands of billions of "plates" and "stellars" did not resemble each other in the least. Exactly what did these scientists refer to? And why couldn't they find the cause for this magnificent phenomenon?

☞ Q 110.

It is considered one of the greatest scientific works of the twentieth century, although the "Axiom of Reducibility," which was part of the "Theory of Types," has never obtained general acceptance among scientists and caused one of the two writers to revise part of the above thesis a dozen years later. Who were the writers of this magnum opus, which among other things showed that logic and mathematics were one unique discipline? Hint: They tried to prove that almost all mathematics could be deduced from a comparatively few "primitive propositions" of logic.

☞ Q 111.

It has been maintained that the Turks, Persians, and Arabs did not have a word for it at first, but frequently referred to it by its Italian name *bùssola.* Others assert that it originated independently in Europe and in China at the end of the eleventh century, although the first documented mention of it was in 1086 in a series of essays called *Dream Pool,* by the Chinese scientist Shen Kua. It's as necessary to international trade and travel as anything in the last nine centuries. What is it?

☞ Q 112.

More than two thousand years ago someone formulated the basis for teaching geometry, and his theories are still valid today, although many were based on definitions, theorems, and axioms predating his birth by a good two hundred years. Who was this mathematical genius and what was his main work, consisting of thirteen books?

☞ Q 113.

How would a woodcutter set about cutting a large log into eight identical pieces, using his saw exactly three times?

☞ Q 114.

A historical and scientific question rolled into one: Can it ever be proven scientifically whether the Nazi battleship *Bismarck* was scuttled by its admiral, Günther Lütjens, or whether it was actually sunk by the British Navy on May 27, 1941?

ANSWERS

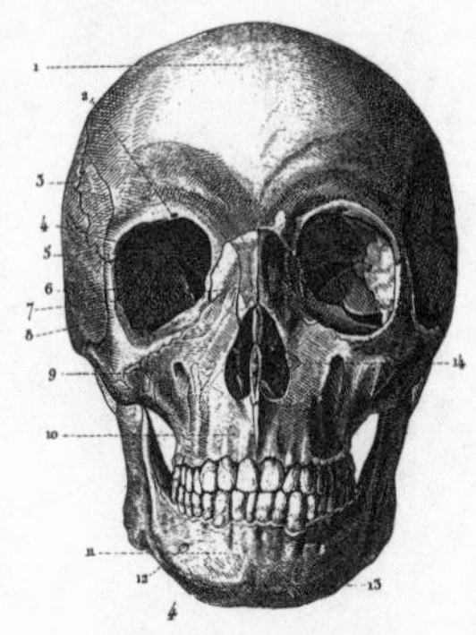

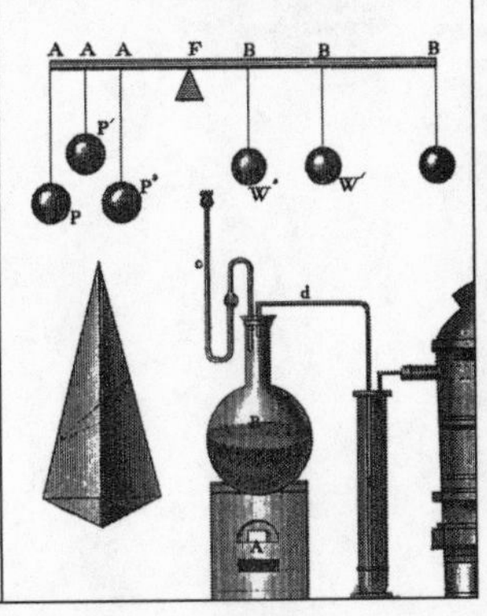

THE NATURAL WORLD

☞ A 1.

Combining these caffeine-rich nuts from the tree *Cola acuminata* with an extract from coca will produce the main ingredients of Coca-Cola. The heart stimulant derives from the oil kolanin. The cola tree is a native of West Africa, but it also grows in Jamaica, Brazil, and India.

☞ A 2.

Worker ants can live up to three years while their queens are known to survive twenty-nine years. Some of the latter are so heavy with eggs, laying up to a thousand a day underground, they are fed their entire lives by worker ants (always females) and never see the light of day. The worker ants can carry loads up to fifty times their own body weight. However, South American and African army ants do not attack armies. They *act* as armies and can consume the weight of an elephant or a crocodile within three days.

☞ A 3.

It is a colorless, fuming, vesicant (blister-producing) liquid acid (HCOOH), also known as formic acid, that is used chiefly in dyeing and finishing textiles. It is now obtained chiefly from oxalic acid distilled with glycerin.

☞ A 4.

Both hold records for discovering most of the comets in their own centuries: France's M. Pons is reputed to have spotted as many as thirty-seven in the nineteenth century, with some of them bearing his name, such as Pons-Winnecke's comet. Arizona's Ms. Shoemaker has already discovered at least twenty-eight comets in the twentieth century.

☞ A 5.

The branch of science is botany.

A *hotbed* is an area of soil enclosed in a low glass frame and heated by fresh, fermenting manure which, as it decomposes, is sufficient for raising early planted seedlings. Today this process is frequently replaced by electric soil-warming cables.

A *chat* can be a small inferior potato, but also the seed of various plants, such as an ament, a spike, or the samara fruit of an ash, maple, or elm.

A *cap* is the umbrella-shaped top of a fungus, such as a toadstool's cap.

Punk is the fungus that grows on rotten wood, a term used more frequently in the United States than in Great Britain.

A *pitcher* is not only a leaf with an appendage shaped like a pitcher, but one that secretes a liquid to lure and digest insects.

A *nucleus* can be a young clove of garlic, a small bulb, the hilum of a starch granule, and in lichens and some fungi, the center of an apothecium or perithecium.

The *mast* is the fruit of beech, chestnut, and some other trees, mainly used for pig and boar fodder.

A *maiden* can be a newly rooted runner, especially when it concerns strawberries, but it can just as well be applied to young fruit trees and to rose bushes when they are being grafted in their first year but prior to their primary pruning. It also refers to soil that has never been plowed.

☞ A 6.

One has nothing to do with the other. The science and art of extracting and reclaiming silver from industrial processes is called silver metallurgy. (More than three-fourths of the world's silver is obtained as a by-product of the refinement of base-metal ores: primarily those of zinc, copper, and lead.) Silviculture is the theory, study, and practice of planting trees for the preservation of forests, for the good of the environment, and for the coincidental production of industrial and domestic goods that will most advantageously meet the demands of society. As a matter of fact, the word *silvan* (sylvan), which means any natural growth pertaining to a wood or forest, is the root of the word *silviculture*.

☞ A 7.

The last truly gigantic meteorite (or 100-foot asteroid, as NASA claimed in 1993) smashed into Earth from outer space less than a century ago, in 1908. After breaking into several chunks in the atmosphere, its biggest part crashed near the Tunguska River in Siberia, leveling the forest there for hundreds of miles. The impact on Earth caused aerial shock waves detected around the globe. The largest crater produced by a meteorite in the United States is near Flagstaff, Arizona. It was caused by a meteorite that weighed about 10,000 tons and collided with the earth about 25- to 50,000 years ago. It is not known, however, how many meteorites have struck Earth from outer space since many of the resultant excavations have filled up with water and become inland lakes, some of them hundreds of thousands of years old. Other meteorites, maybe as many as three out of four, have crashed into our oceans, unbeknownst to us. Nonetheless, we have established scientifically that there are at present circa 130 known meteorite craters on our planet. As a matter of fact, we are bombarded by about 26,000 meteorites every year, but thousands of others burn up in Earth's atmosphere. Those that get through frequently are no larger than 1⁄16 of an inch (micrometeorites). However, it is known that the extraterrestrial weight landing on earth totals several tons per day.

☞ A 8.

Foxglove. One species is the source of the drug digitalis. The ingredients of digitalis that strengthen and help regulate heart contractions are glucosides (vegetable substances that yield glucose when decomposed) that have been isolated in crystalline form from the garden variety foxglove: digitoxin, gitoxin, and gitalin. These are referred to as glucoside digitalin. The word *digitalis* derives from the Latin name for foxglove, *Digitalis purpurea.*

☞ A 9.

If you said none, it wouldn't be literally true. Even though meteors are luminous bodies in the night sky and are more commonly known as shooting stars, about 11,000 tons of meteors and micrometeorites annually manage *not* to burn up completely in the atmosphere; they land on Earth as fragments and dust particles. Their mineral composition is believed to be more fragile than that of meterorites to which they probably are related.

☞ A 10.

It is the budgerigar, a parakeet that is commonly light green with black and yellow markings, although it is bred under domestication in many colors. Among the many names by which the budgerigar has been known are warbling grass parakeet, undulated grass parakeet, scallop parrot, shell parrot, canary parrot, and zebra grass parakeet. This lovebird is a native of Australia.

☞ A 11.

Seashells. The Philippines are the largest exporters of seashells, shipping between one and two thousand tons per year, while Haiti, Kenya, Indonesia, Mexico, and India also export seashells primarily to the United States, Europe, and Japan.

☞ A 12.

On the contrary! The shield of the much-abused ozone layer actually develops when the sun's ultraviolet radiation strikes the oxygen shield around our planet in the stratosphere. The ordinary oxygen molecules that exist primarily six to thirty miles above Earth are converted from two-atom molecules to three-atom molecules by the sun's ultraviolet radiation, forming the safety belt around the globe called the ozone layer. This is an allotropic form of oxygen, which protects the Earth from the sun's ultraviolet rays. This slightly pungent gaseous layer, in fact, is now being depleted not by the sun's radiation but by man's release of synthetic chemicals into the atmosphere. The ozone layer protects life on this planet while mankind's chemicals, such as CFCs, used in air conditioning as well as in the manufacture of plastics and a result of the burning of forests, will stray into the stratosphere. There the sun's ultraviolet radiation disintegrates the chemicals' molecules, converting them into unfettered chlorine atoms that have already damaged and will continue to ravage and vaporize Earth's protective layer.

☞ A 13.

Lampreys and sharks have skeletal structures made up entirely of cartilage. They have no bones. Henry I of England is believed to have died from eating too much of his favorite dish, lamprey, which is difficult to digest.

☞ A 14.

Technically, from the astronomer's point of view, this is not quite correct. The dense mass of matter itself did not cause the explosion, known as the Big Bang, or the beginning of time and the universe, but rather the substance of space itself began to expand and with it the original fabric of the physical universe, "transporting" all the newly formed galaxies into space. The continuous expansion of space spreads the known matter of the original mass of the universe farther and farther away from its point of origin. The physical properties of the galaxies themselves hardly expand, though, only the space between them does.

☞ A 15.

Kale and collard.

☞ A 16.

They're all gastroliths, which means that they swallow stones to help their systems to break up and digest food.

☞ A 17.

It is true that astronomers don't know too much about black holes yet, but what they do know is that if any companion star, for example, gets too close to what is believed to be a black hole, the star is literally "sucked in." There is a sudden incredibly bright flare-up, and then the total extinction of the ensnared star's light. It has literally been swallowed up. Even the gases of passing stars, too far from the black holes to be ingested by them, are drawn in.

☞ A 18.

Lore is the space between a bird's eyes and its bill (the same term is used with serpents). A zygodactyl's circumference is measured and determined by the bird's characteristic two forward toes and two rearward toes.

☞ A 19.

Although Mother Earth is targeted by only 2.2 billionths of the power generated by the sun, the actual amount of energy that the sunlight supplies to the Earth's atmosphere is about two calories per square centimeter per minute. This amount is known as the solar constant. As far as we know, it is atomic transmutation (the process by which one element in the periodic table is converted into another substance or condition) that generates the energy needed to sustain the sun's storehouse of power.

☞ A 20.

By studying the rings of ancient trees. Unlike anything else in nature, trees react permanently to changes in temperature. The barometer of the various seasons, whether marked by downpours or other environmental conditions such as sudden frosts or long spells of drought, is reflected in tree rings. Good summery weather with the right amount of rainfall will let the annual ring be reasonably fat, while in periods of drought or sudden cold spells the annual ring will be correspondingly narrow. Both summer and spring rings are composed of different widths, formations, and shading within each annual ring. Thus, tree expert Lisa J. Graumlich could learn from her study of western junipers and foxtail pine trees in the Sierra Nevada that North American trees between A.D. 1100 and 1375 went through an era of Mediterranean-like weather. This climatic change was so widespread for almost 300 years that vineyards abounded in Britain and the Vikings farmed land in Greenland that is now frozen and barren, a result of the Little Ice Age from 1450 to 1850.

☞ A 21.

Only birds do not develop cataracts.

☞ A 22.

No, they're not the nymphs inhabiting the sea, rivers, or forests mentioned in heroic tales. These nymphs are the kind of young insects that undergo incomplete metamorphosis. One kind of nymph that can regenerate amputated legs is the baby cockroach.

☞ A 23.

There is no difference. Since their size varies between one and several hundred miles in diameter, they are not big enough to be referred to as planets, but more than 1,600 are large enough to have been spotted. Most circle the sun between the orbits of Mars and Jupiter. The best known among them is called Juno, and it is believed that these heavenly objects resulted from collisions and adherence of small bits of matter, rather than that they solidified and grew into a mass, as did the planets. Asteroids remain in the same orbit due to the enormous gravitational force of nearby Jupiter, which is so tremendous in size that the other eight planets of the solar system would fit more than twice inside this giant. This huge, bitterly cold planet is not solid, like Earth, but mostly composed of hydrogen and helium, and, like the sun, is liquefied by pressure in its interior.

☞ A 24.

They are the four largest moons circling Jupiter. Io is not only the one closest to Jupiter but probably the most volcanic celestial body in the solar system. Jupiter, located 482 million miles from the sun, is the largest planet in the solar system.

☞ A 25.

The end product is one of the world's favorite beverages: tea. It is assumed that tea leaves were first brewed in the court of Chinese emperor Shen-mung about 2500 B.C. Although tea has been considered England's national drink for centuries, Shakespeare never mentioned the brew in any of his plays.

☞ A 26.

The sea horse. The female places eggs for hatching into the male's brood pouch on its abdomen. After an incubation period of forty to fifty days, as many as 200 young are delivered from the male's pouch.

☞ A 27.

Timothy is one of the most important hay grasses. It is found in areas that are relatively cool, temperate and humid, and is named after Timothy Hanson, an American, who took the seed from New York to the Carolinas, c. 1720. The plant (*Phleum pratense,* of the order *Cyperales*) grows well in combination with red clover and alfalfa. This valuable fodder grass predominates in the northern half of the United States, where timothy grass crops are rotated with pasture and hay seedings. In the 1990s, however, bromegrass tends to take the place of timothy mixtures in some of those parts, and generally it is now more appropriate for the use of hay than for pasturage.

☞ A 28.

The largest star in the constellation *Orion* is Betelgeuse. This giant red star is believed to be between 300 and 500 times the size of the sun (about 200,000,000 miles in diameter). It is so large that the orbits of Earth, Mercury, Venus, and Mars would fit inside. It is also a much cooler star than our sun. The sun is considered a medium-hot star that burns at about 11,000° F (c. 6,000° C). Betelgeuse, which is about 650 light years away, burns at about 5,000° F (c. 2,800° C).

☞ A 29.

A selenologist is a scientist who studies the moon. Tycho is a comparatively young crater on the moon, created by the impact of a giant meteorite a few hundred million years ago. Tycho can be seen on the moon's southern hemisphere with a pair of binoculars during a full moon. The terminator, within the confines of this question, is the frontier that separates the dark and illuminated parts of the moon.

☞ A 30.

To start with, cocoa is a tropical palm tree, *Cocos nucifera.* The cocoa nut (sometimes spelled coconut) is its fruit, a rough, hard-shelled nut containing an edible white lining inside the fruit and a sweet liquid known as coconut milk. However, this fruit should not be confused with the other cocoa, which is a drink prepared from cocoa beans, the seeds of *Theobroma cacao.*

Coconut butter is a solid oil obtained from the white lining of the coconut, while cocoa butter is the substance extracted from the cacao nut in the manufacture of cocoa (from which the cacao drink is finally derived). Confusing? As long as you add hot milk to the cocoa powder from cacao seeds that have been roasted and ground you'll be fine. And for solid food you can have some cocoa-nut slices with it.

☞ A 31.

The watermelon, the edible fruit of *Citrullus lanatus* or *vulgaris.* It can grow to be 38 kilograms for commercial use, although the 1990 world record for the largest watermelon was 262 pounds. The main body of most melons is comprised of about 91 percent water, while the sugar content can reach 12 percent.

☞ A 32.

The Greeks labeled the constellations in their own tongue because it was the northern sky, containing these constellations, that was observable from their empire. Most professional astronomers, however, scoff at the notion of reading peoples' characters or future in zodiacal signs. In principle, many astrologers maintain that the universal aspect of the stars' and planets' gravitational force exerts an influence on children at birth. Newton (1642–1727) already proved that everything exerts a force on everything else; we can expand on Newton to include the father's sperm, the mother's ovum, hereditary factors, the obstretician's treatment, the food consumed, the economical, familial, and natural surroundings of the mother and the newborn child. So, to some extent, the astrologers are right. However, ancient zodiacal charts are frequently used today, even though the position of Earth has changed since the time of the ancient Greeks and the old zodiacal signs no longer accurately represent today's position of the sun and Earth. Different astrologers will give different analytical interpretations for this reason, yet they will trust their own interpretation of the constellations, and the astronomers' refutation of them plays no part in their belief.

☞ A 33.

The scientist who made these observations was the painter Leonardo da Vinci (1452–1519). He also observed that the the year's moistness can be determined by the width between the tree's rings. The farther apart the rings, the more moisture there was generally in the ground surrounding the tree.

☞ A 34.

The order of beetles and weevils (*Coleoptera*), with about 290,000 known species. Since the end of World War II more than 40,000 new species have been detected.

☞ A 35.

Yes, astrophysicists claim to know what happened: Matter was just about to exist in the form of elementary particles. Before that, the weak forces of the universe had coalesced with the electromagnetic force, which occurred about ten-billionths of a second after the Big Bang. Matter existed in the entities called quarks and leptons—elementary particles such as muons and electrons. After that, neutrons, protons, and other particles of the most elementary nature were formed.

☞ A 36.

Gnats, mosquitoes, flies, and midges. In the last thirty years, more than 10,000 new species of the order *Diptera* have been located and classified.

☞ A 37.

Nobody knows for sure why the corona is hotter than the rim of the sun. There are many theories to explain this puzzle, but the most frequently quoted is that the enormous heat of the corona is due to the release of energy from magnetic fields. Would you be cremated in a fraction of a second inside the region of the corona? Not at all. Since the atoms in the corona are scantily distributed and have been divested of most of their electrons, the energy making up its heat is so spread out that, surprisingly, a human being could stand inside the corona and never feel the high temperatures.

☞ A 38.

Born Carolus Linnaeus, in 1761 he was granted a patent of nobility and henceforth called Carl von Linné (1707–1778). His taxonomic system, more than that of any other scientist, enabled botanists to place a plant swiftly in a named category. *Systema Naturae* (1735) catalogues about 4,200 animals, and *Genera Plantarum* appeared two years later. The fifth edition of the latter work, published in 1754, describing thousands of plants, was internationally recognized as the definitive naturalist work of that period. Yet his most lasting contribution to science was his system of "binomial nomenclature," in which each plant and animal organism is represented by a two-word Latin or Latinized name. The first name indicates genus (e.g., *Homo*), and the second indicates species (e.g., *sapiens*). *Homo sapiens,* of course, is the designation given to Thinking or Modern Man.

☞ A 39.

The North American periodical cicada, belonging to the order *Hemiptera,* has a metamorphic period of seventeen years to grow into an adult. Often mistakenly referred to as the seventeen-year "locust," this cicada lives in the eastern United States. The southeastern cicada requires thirteen years to grow into an adult.

☞ A 40.

Pluto's rotation carries it inside the orbit of Neptune, but it does so for only part of the Plutonic "year," which lasts 248.5 Earth years as it orbits the sun. Pluto is 3,666 million miles (5,902 million km) away from the sun and was discovered by Clyde Tombaugh in 1930. In fact, the planet is so small—1,519 miles in diameter, 0.5 percent of the volume of Earth, and its mass about 0.001 that of the Earth—that some astronomers believe it to be a captured comet, although it does have its own moon, Charon, which was discovered in 1977. Pluto's orbit is highly eccentric: In 1989 it came closer to the sun than Neptune did, and the orbits of Neptune and Pluto do not now intersect. It is about forty times as far away from the sun as Earth and not much is known about the planet, so that most calculations regarding Pluto are still imprecise.

Charon is not be confused with the asteroid Chiron, which moves in an orbit between Saturn and Uranus. Both Charon and Chiron were discovered in 1977. Lately, Chiron has developed a nebulous envelope, like a comet.

☞ A 41.

There are two more reasons. First the more obvious one: chemicals to retard spoilage have been reduced for environmental and health reasons. More important, though, the metabolism of the peach burns up its sugar reserves faster since it "breathes" more swiftly and in a chain reaction consumes more of its sugar deposits. Sunlight and warmth penetrate the peach's fragile epidermis more quickly, and the higher the temperature, the faster the peach burns up its sugar reserves; it "sweats," and spoilage results.

☞ A 42.

Today they are known by these names: a) yellow-shafted flickers, b) shrikes, c) bobolinks, d) peregrine falcons, e) mourning doves, and f) sage grouses.

☞ A 43.

About 99 percent of the gigantic cloud of interstellar dust formed the sun and only about 1 percent of it ended up in the formation of all the planets and other matter in the solar system. The gyration of the dust cloud spun the evolving planets into a flat orbital disk, the ecliptic. With the exception of Pluto, all of the planets in our solar system reside on this plane and move clockwise when observed from above the Northern Hemisphere.

☞ A 44.

Venus, the planet nearest to us, rotates once every 243 days (from east to west, like Uranus), and passes between the Earth and the Sun every 584 days. What has perplexed astronomers is why we always see the same face of Venus whenever it passes Earth. Jupiter, which has a mass 318 times that of the Earth's, with a volume 1,300 times that of our planet, makes one complete revolution on its axis about every 9 hours, 51 minutes. The strangest feature about this planet is that it contains the Great Red Spot that is about 8,000 miles wide and 20,000 miles long. It is believed to be the center of an incredibly destructive hurricane that has been in existence since around the twelfth century. Jupiter takes 11.86 years to orbit the Sun, has at least sixteen moons (one of them—Io—has active volcanoes) and is composed largely of helium and hydrogen.

☞ A 45.

The ecological disaster Dr. MacDonald refers to is the almost total destruction of the American chestnut tree.

Since 1904 a bright orange fungus has destroyed hundreds of millions of mature chestnut trees in the United States alone after trees imported from the Far East at the beginning of the century infected them with their blight. In the 1990s Dr. MacDonald, perhaps the leading chestnut expert, has been in touch with two molecular biologists, Dr. Gil H. Choi and Dr. Donald L. Nuss, and they hope to "treat" the chestnuts with a virulent, but *less* virulent, genetically altered strain of Cryptogamia, or fungus, that will triumph over the much more virulent strain of fungi affecting the chestnuts today. This, in turn, could save the young chestnuts that are still sprouts at the moment and cover the forest floors of several Eastern states. However, it will take several years before the scientist will be permitted to even release the less virulent fungus in some of the affected forests. Other biologists and geneticists, like Dr. Mark McClure, Dr. Charles R. Burnham, and Dr. Sandra Anagnostakis, are trying to stem other blights besieging trees—deadly fungi that infect the American elm now as well as the butternut tree, the Eastern hemlock, and the white ash, putting all these trees in danger of extinction early next century.

☞ A 46.

This bird is popularly known as the bald eagle—the national bird, and emblem, of the United States. It lives in North America and Canada. The adult bird has a fully feathered white head, neck, and tail, and its correct common name among ornithologists is the white-headed sea eagle.

☞ A 47.

Virtually all fish are cold-blooded animals, except tuna. A big-eye tuna, for example, can change the rate that it loses or gains warmth—heat conductivity—by as much as one-hundred-fold in less than one minute. All species of tuna have the ability to exercise their heat-generating muscles and to regulate their body temperature. This allows them to conserve heat in their muscles and to remain relatively warm in cold water. To seek food in the cold depths of the ocean, the tuna "switches on" its heat exchangers, then turns them off as it returns to warmer surface layers of the water. There are about two dozen more fish species that are warm-blooded, but the difference between them and tuna is that the others can deploy their heat-generating muscles only in parts of their bodies, such as their brains and around their eyes. However, all of these heat-generating fish, including tuna, the butterfly mackerel, marlin, swordfish, and billfish, belong to the suborder *Scombroidea.*

☞ A 48.

The planet is Venus, and when it shines brightly in the east at dawn it is also known as the Morning Star. It is referred to as an *inferior planet,* like Mercury, because there are years when it orbits closer to the sun than Earth does. Planets orbiting beyond the Earth are called the *superior planets.*

☞ A 49.

Most of us take the moon so much for granted that we are not aware that it is virtually invisible to the naked eye once a month. This relatively short phase is called *new moon,* and it occurs when the moon hovers between Earth and the sun. The part of the moon lighted by the sun is facing away from Earth which faces the unlit half of the moon. Shortly after the moon's "disappearance," it makes its comeback as a thin crescent curving toward the right.

☞ A 50.

There's no reason to complain. The "dolphin" we are served in restaurants is a fish with gills and iridescent colors. In Hawaiian restaurants it is often advertised as mahi-mahi. The intelligent *Flipper*-style dolphin is one of about three dozen species of cetacean mammals in the family *Delphinidae.* In contrast to the fish that can reach five feet in length and is cultured for the restaurant market, the dolphin we love is a bottle-nosed, air-breathing mammal who, like Orca the killer whale, must return to the surface of the water to inhale. The edible dolphin fish (*Coryphaena hippurus*) extract oxygen from the water they swim in like any other fish.

☞ A 51.

A dendrochronologist is a scientist who studies the age of trees. The hardest thing he or she may have to decipher is the distinction between genuine and false tree rings. Besides the annual rings, visible when a tree is cut, some giants (or dwarfs) of the forest also display false intra-annual rings, which may have been caused by a midyear cold snap.

☞ A 52.

It's not science fiction that the six-mile-long comet Swift-Tuttle may come perilously close to Earth in 3044, but not in 2126, as first predicted, according to Dr. Brian

G. Marsden of the Harvard-Smithsonian Center for Astrophysics in Cambridge, Massachusetts. Despite the comet's passage through the inner solar system every 130 years or so, it had not been sighted between A.D. 188 and 1747 because it was too far from Earth to be visible with primitive telescopes. This comet is also responsible for the annual Perseid meteor shower. Its shedding of dust and debris forms a stream of particles that gives the appearance of coming alive with flashes of light. The Perseids are the best known meteor showers in any given year. However, as far as running out of carbon dioxide is concerned, we do not have to worry. Scientists at Pennsylvania State University concluded in the December 1992 issue of *Nature* that this sorry event will not occur a hundred million years from now, as forecast earlier, but that the decline of carbon dioxide in the atmosphere, causing plant life to be starved of its main chemical building blocks, will not occur for at least another billion years.

☞ A 53.

It is neither. The edible, juicy part really is a transformed part of the stem, and this red, fleshy receptacle bears the achenes on its exterior. These small, hard, dry, one-seeded indehiscent pips (achenes) are actually the seeds. *Indehiscent* means that these seeds do not open at maturity. However, the plant in its entirety—the genus *Fragaria*—is a fruit.

☞ A 54.

Scientists believe that the brown dwarfs have been integrated in their community. Actually, brown dwarfs are objects less massive than stars but larger than planets. They are part of the community of orbiting bodies circling huge stars. At present it is believed that brown dwarfs outside the solar system are extremely large gaseous bodies, even larger than Jupiter, and that they were stars that failed to evolve, yet are closer in mass to the planet Jupiter than to stars.

☞ A 55.

You should agree with the statement, because we are referring to nitrogen fixers in plants. Nitrogen molecules (N_2) taken from the air for the sustenance of plant life have to be chemically transformed (or "fixed") into ammonia (NH_3). Multicelled plants could not exist without this nitrogen fix, and therefore life would cease, and neither animals nor human beings could survive. Virtually all this transforming or transference (fixing) is done by bacteria that are present almost exclusively in nodules of the roots of leguminous plants, and the resulting nitrogenous compounds are made available to their host plants. In bodies of water, the nitrogen is transformed (fixed) by algae and bacteria.

☞ A 56.

Many snakes can generate a general degree of warmth from the air and ground around them, but a python female appears to be the only snake that can twitch and move her body so quickly that the friction generates heat. She then surrounds her eggs closely with her body to warm them.

☞ A 57.

Both theories are inaccurate. The geocentric theory claimed that the Earth was at the center of the universe, with the sun and moon circling it. Copernicus proved that the Earth, moon, and stars revolved around the sun and therefore he believed, mistakenly, that the sun was the center of the universe (the heliocentric theory).

☞ A 58.

The flower, *Rafflesia arnoldi,* is named after the founder of Singapore, Sir Thomas Stamford Raffles (1781–1826) because he was a great patron of science. This stemless parasitic plant, found in Java and Sumatra, unfortunately has no horticultural potential. Its huge dioecious flowers have a calyx of five fleshy lobes that emit an odor resembling that of carrion. They can measure up to three feet across and weigh as much as 15 pounds.

☞ A 59.

None, according to Fred Koontz, curator of mammals at the New York Zoological Garden. Although most animals with eyes produce tears, none of them do so for emotional reasons but only to protect and cleanse their eyes to keep them in good working condition.

☞ A 60.

It took more than 170,000 years, perhaps as much as 200,000 years, at the speed of 186,282 miles (299,792.8

km) per second, for the spectacular burst of light of the LMC supernova to reach the Earth. (In contrast, the Moon's light takes a second and a third to reach Earth.) The Magellan's actual distance from our planet is over a billion billion miles, and the light rays of the Little and Large Magellanic Cloud (their luminosity) have a brilliance about 800 million times that of the sun.

☞ A 61.

The first is that it is the only bird that can fly upside down. The second is that it is the smallest bird in the world. Hummingbirds are found only in the Americas. They fly upside down for part of their in-flight maneuver, but they do not do so for long and they can only do it because the structure of their wings is angled. They can also fly backward, in order remove their bills from probing tube flowers for tiny insects and nectar. The fairy or Princess Helen's hummingbird of Cuba, whose body is two inches long, (*Calypte helenae*) and the bee hummingbird (*Mellisuga helenae*), with the adult males measuring 2.24 inches in total length and also found in Cuba and the Isle of Pines, are the two smallest birds in the world.

☞ A 62.

It certainly is not the mosquito's intention to cause an itch. The enzyme of the mosquito's saliva when the insect bites us has only one purpose: to prevent the victim's blood from clotting, thus making it indigestible for the mosquito. Once the insect's proboscis has penetrated our skin, the mosquito's enzyme prompts our immune system to release histamine. This histamine causes the itching, swelling, and rash. Water-softened meat tenderizer containing papain (which acts to break down protein molecules), when applied to the site of the bite, will stop the itch in no time.

☞ A 63.

Today we still see the gaseous remnants of the Crab Nebula, which was first observed in 1054. It blazed into view on July 4 of that year in the constellation Taurus, near Orion. This explosion occurred only 6,000 light years away. Light travels about 5,880,000,000,000 miles in one solar year.

☞ A 64.

It's the field mouse (*Apodemus sylvaticus* and *Microlus pennsylvaniacus*). When scurrying through its tiny pathways, it feeds on any obstructing plants and their roots to keep its trails clear and tidy. Of course, the motion picture "star" based on its likeness is Mickey Mouse.

☞ A 65.

There will again be two lunar eclipses and five solar eclipses in 2160.

☞ A 66.

The man responsible for the superb illustrations and descriptive accounts of the life cycle of these beasts was Robert Hooke (1635–1703). The genus *Pulex* he described with so much loving care is better known to us as the flea, the *Culex* is the bothersome gnat, and the *Pediculus* is the awful parasite better known as the louse. The otherwise little-known Hooke also invented the marine barometer and sea gauge among other things, and Hooke's Law stipulates that the tension in a lightly stretched spring is proportional to its extension from its natural length.

☞ A 67.

The Andromeda galaxy is more than 2,200,000 light years from our planet. The Little Magellanic Cloud supernova, first sighted in 1885 and still visible in the night sky, is a mere 170- to 200,000 light years from Earth.

☞ A 68.

The plant's name is mandrake (*Mandragora afficinarum*). The alleged habit of this much-maligned plant is that, as Shakespeare put it in *Romeo and Juliet,* it "shrieks like mandrake torn out of the earth . . ." At one time its roots were also used to promote conception and it is supposed to have the properties of a soporific and cathartic. It is most commonly found in southern Europe and northern Africa.

☞ A 69.

Cows sure look as though they are ruminating or thinking about something. But in their case, ruminating is not a matter of a thinking process but of a digestive procedure. The answer is that while they are ruminating *and* chewing the cud, the chewed-up grass does not go to the cow's stomach but to an antechamber called the rumen (from which the word *ruminate* derives) and it is here that the grass is dissolved by microorganisms. The partly broken down grass that remains is returned to the animal's mouth to be champed again, as the fully digested grass and microorganisms are being secreted through the real stomach as fatty acids. The udders then store the liquefied part of the cattle's nourishment, which is composed of 87 percent water (for cows' milk), 3.5 percent protein, 3.9 percent fat, 4.9 percent milk sugar, 0.7 percent ash, carbohydrate, and minerals. Sunlight provides cows with an adequate supply of vitamin D to utilize the calcium in their feed. This vitamin, mixed with calcium, is passed on to milk drinkers.

☞ A 70.

The Latin name of the flower, Rangoon Creeper, is *quisqualis.* The botanist Rumphius (1627–1702) first detected the plant in Malaysia in the seventeenth century. While the Malay name for it is *Udani* and means "how—what!" after it had been rendered in Dutch as *hoedanig,* Rumphius translated it into Latin as *quis qualis,* registering his surprise at the plant's structure, and this curious play on words stuck.

☞ A 71.

It is not any star in particular—but that light comes from all the stars in space that are visible to the naked eye.

☞ A 72.

The planet Earth, as it moves approximately 93 million miles from the sun in an orbit of close to 600 million miles.

☞ A 73.

The sunlight hits the poles at such a shallow angle that the bottom of some craters have lain in shadow for billions of years. Since there is no atmosphere to generate a semblance of heat in these "subterranean" spots, the temperature there is far below zero (–170° C). The ice that condensed as frost in these craters, therefore, is considered to be still there—frozen, rock-hard and unchanged.

☞ A 74.

All plant life. The carbon dioxide is taken from carbon and oxygen atoms in the air, and the minerals from the soil surrounding the roots, which then convert these materials into tissue that changes inorganic into organic or living matter.

☞ A 75.

If we were referring to nuclear tests, they would blow Mother Earth to smithereens. The explosions concern the second most powerful kind of stellar force known in the universe. (It's exceeded only by supernovas.) Between twenty-five and seventy-five detectable novas explode in the Milky Way Galaxy each Earth year, and 10^{38} joules of electromagnetic radiation is about equal to the energy radiated by the sun in 10,000 years. Only about one such binary explosive force is visible to the naked eye every ten to twenty years, although the general public has not witnessed such a cataclysmic event since 1975, when Nova Cygni became the brightest focal point in its constellation for about forty-eight hours.

☞ A 76.

One of the most economically important and dominant hardwood trees: the American chestnut tree, especially the one found in Appalachia.

☞ A 77.

It is not a drug at all but *spirulina,* which has been recognized around the world as the most vital and promising of all dietary supplements produced from microalgae. This blue-green plant organism is created by the interaction of sunlight and water (photosynthesis). It thrives anywhere in alkaline waters and sunny climates, promising a breakthrough in world food production.

☞ A 78.

All are located around the sun. The rim of the sun, which would be comparable to a luminous envelope around the star, is the photosphere. The two layers above the photosphere are the chromosphere, which is the gaseous envelope through which the sunlight passes from the photosphere, and the corona. The latter is also the glow you see around the moon during a total eclipse of the sun. Solar flares, sunspots, and eruptions of incandescent gas from the surface of the sun (prominences) are normal solar activities in the chromosphere and photosphere.

☞ A 79.

The grunion, a member of the silverside family of the southern California coast, is the only fish known to come inshore to spawn on wet sand when the moon is nearly full.

☞ A 80.

These were the elements created at the moment of the Big Bang. All other chemical and physical properties were created later. But *later* in this case can mean a billionth of a second later.

☞ A 81.

The Greek word, derived from the Latin for "long hair," is *kometes,* and our English word *comet* can be traced to it. The comet's body, including its tail, can be extremely long—up to 150 million miles—and its curved tail is made of dust particles. Two-tailed comets sport a second tail made of glowing gas. The closer the comet's body gets to the sun, the more it is drawn out and forms this long, bushy tail, which really is the comet's melted surface ice. But the tail will always point away from the sun because the solar wind blows the gas and dust that has formed a hazy cloud (coma) away from the wind's source (the sun).

☞ A 82.

The sun is about 330,000 times the mass of the Earth, although it is "only" about 864,000 miles in diameter (109 times the diameter of the Earth). It is a rather small star, comparatively speaking, with the next-nearest star almost 300,000 times as far away, if you consider that our approximate distance from the sun is 93 million miles. The two most prominent gaseous chemical elements of the sun are hydrogen (69.5 percent) and helium (28 percent). Another five billion years down the road (the approximate age it is right now) the sun will become hotter and hotter due to the increasing deposit of the helium and its ash at the core, and the heat will ignite the unfused hydrogen gas surrounding the core. As a result, the surface layer of the sun will be pushed outward toward Earth, Venus, and Mercury, finally engulfing them and incinerating life in all its manifestations, especially on our planet. Oceans will boil and bubble and be converted to steam, with the sun finally collapsing into itself and becoming a white dwarf star with a temperature of about 120,000° C. Perhaps a billion years later it will become a cold black dwarf star. Earth will have shriveled into an eternally silent orb (Shakespeare's "the little O"), frozen beyond recognition.

☞ A 83.

What these American scientists shared was the color of their skin: all of them were black. Rilleux was the inventor of the vacuum pan. Woods invented the third-rail system for trains. Blair was the inventor of corn- and cotton-planters. Dr. Hinton developed the Hinton tests for syphilis detection. Dr. Just was a famous marine biologist. Dr. Drew was the blood bank pioneer. Dr. Williams was a surgeon who helped perform the world's first successful open-heart surgery. Dr. Julian developed commercial production facilities for cortisone.

☞ A 84.

False. Earth's magnetic field has the tendency to alter its course now and then, and if you have the answer to what causes this planet's magnetic field's erratic nature, you may be a candidate for a Nobel Prize in physics. The occasional change in direction in our magnetic field still puzzles geophysicists and astronomers. As a matter of fact, the sun's magnetic field is known to be similarly affected, and its magnetic behavior is a mystery to astronomers as well. The sun's magnetic field is believed to reverse itself about once every eleven years, in apparent synchrony with the eleven-year cycles of sunspots.

☞ A 85.

Literally translated, a centipede is an animal with a hundred feet. But virtually no centipede has exactly one hundred legs. Centipedes can have anywhere between thirty and 346 legs, just as millipedes don't have exactly a thousand feet, but anywhere between forty and 800 legs. Arthropods (often spelled arthropoda) are the largest phylum of the animal kingdom, boasting close to 800,000 species, including insects, crustaceans, and arachnids—about four out of five of all the known species of the animal world.

☞ A 86.

The energy is sunlight, and the entire process of plant growth results from photosynthesis, without which no life on Earth could exist. Plants supply other living things with oxygen and carbohydrates; but first the energy inherent in sunlight must synthesize other carbohydrates and sugars in plants.

☞ A 87.

The answer to the first question is quite simple: Nobody knows. We do know that the galaxies are still moving away from each other, meaning the universe is still expanding. The American astronomer Edwin Hubble (1889–1953) discovered this in 1923, and six years later he found out that velocities of nebulae (red shifts) increase in a linear manner with distance, meaning that in far-away galaxies the wavelength of the light was longer than previously determined. Hubble interpreted this to mean the galaxies were moving away from each other, and the faster they retreated, the farther away they became—approximately 50–100 km/30–60 miles per second for each million parsecs of distance. Moving in the opposite direction, Hubble realized that the universe shrank toward the beginning of time to a small, superdense point maybe 12, 15, or 20 billion years ago. The expansion of this hot, dense state culminated in the Big Bang.

Hubble did not baptize this beginning as the Big Bang. That was left to British astrophysicist Fred Hoyle (b. 1915). He coined this term as a kind of a jest in 1950, when there were still many theories about how the universe was born. Just as puzzling as the the mystery of what happened before the Big Bang is the question of what the expanding universe will grow into in the future.

☞ A 88.

These long-nosed, colorful fish, which do not resemble butterflies, can be found in the Philippines (*Forcipiger longirostris*) and on the Great Barrier Reef (*Chelmon ros-*

tratus). They have dark spots on their tails that look very much like eyes. These spots confuse predators, which tend to attack the wrong end of the fish. Butterfly fish are also among the flattest, thinnest creatures in the fish world; their form allows them to hide under rocks and in the tightest fissures and crannies unnavigable to other fish.

☞ A 89.

Kepler's second law showed that the planets did not orbit the sun in circles but in perfect ellipses. Copernicus (1473–1543) even earlier discovered that Earth revolved annually around the sun, although he thought the orbit was circular, not elliptical. Kepler also discovered that the speed of a planet decreases as it retreats from the sun, and that it increases as it gets closer to the sun. Sir Isaac Newton (1642–1727) derived his own theories of universal gravitation in 1687 by basing some of his work on Kepler's data.

☞ A 90.

You'll find them in woody plants. Xylem is the thin cellular tissue in the woody part of plants that contains and transports dissolved materials, such as moisture. For instance, in a sawed-off branch, you will find a thick, fibrous region with light and dark rings: the xylem. Phloem is the softer, cellular portion of fibrovascular tissue in trees. It has a laminated appearance and is closely connected with the bark. Containing the end products of photosynthesis, the cell tissue of phloem serves as a pathway for the distribution of food material to the woody parts of the plant world.

☞ A 91.

Although sunspots appear dark, they are neither black nor dark brown. They are extremely bright with their luminous heat. They only appear to be darker, redder, even blacker, relative to their lighter surroundings. The reason for this is that these spots, which are a consequence of magnetic storms, only appear in the outer layer, or photosphere, of the sun, and they happen to be cooler, though still incandescent, than their brighter surrounding gases. You will see more of them at the end than at the beginning of their eleven-year cycle.

☞ A 92.

Foot-and-mouth disease has destroyed entire herds of cattle, goats, sheep, and pigs. It is easily transmittable to humans, although pasteurization and the boiling of infected farm products and utensils can prevent further spread of the disease. Some successful treatments in less serious epizootics (animal diseases) can be affected by the use of attenuated-virus vaccines.

☞ A 93.

The flies that infested these wounds are called blowflies (family *Calliphoridae*), and what they actually do is to consume the decaying tissue in wounds. Even more scientifically astonishing is the fact that they produce a substance that destroys harmful bacteria, and this chemical also induces the growth of new tissue. Long after World War I, scientists finally were able to isolate this chemical in labs, and today many medical products have replaced the chemical substance first "introduced" by the much-maligned blowfly.

☞ A 94.

Algol, also known as Beta Persei, is an "eclipsing binary," a pair of rotating stars. In Algol, one star passes in back of the other every sixty-nine hours, causing its brightness to drop by two-thirds. When the smaller star, the brighter of the two, is eclipsed by the larger one, the light that manages to reach Earth from this pair (binary) wavers slightly and gives the impression that it's winking.

☞ A 95.

Astronomy. A Cepheid is a class of pulsating stars whose intrinsic light variations are quite regular. The period/luminosity relationship applies to the length of the period of light variations and the actual luminosity (the intrinsic brightness) of the stars: the more luminous the star, the longer the period of light fluctuations. This provides astronomers with a method of calculating stellar distances inherent in our galaxy and in those outside the Milky Way. By measuring the light-variation period of a Cepheid, astronomers can deduce its intrinsic brightness. When this brightness is compared with its observed luminosity, it is possible to determine the distance. In contrast to the two winking stars that were the subject of the previous question, the Cepheid fluctuations are the cause of intense, short-lived pulsations in the star itself.

☞ A 96.

Indeed it is. Because Jack Dempsey is the only fish named after a boxer. The fish (*Dichlasoma octofasciatum*) earned the boxer's name because it is so aggressive (certainly not recommended for community tanks). Belonging to the *Cichlidae* family, this fish originated from the Middle Amazon in South America and is relatively easy to breed in captivity. The Jack Dempsey's color is a mottled brown with bright blue spots, and it is not known why older males develop a large bulge on their heads or what its function is.

☞ A 97.

Nobody has ever died from consuming it in reasonable amounts. This is the acid that makes unripe apples sour. As the fruit ripens, the amount of malic acid in it declines, and the apple sweetens in the process. Climate and soil also influence an apple's tartness, and some people prefer the somewhat sour taste of green apples like Granny Smiths to the sweeter flavor of other popular varieties. Malic acid is also found in rhubarb and grapes.

☞ A 98.

None whatsoever. The smilodon is a New World genus of the Pleistocene saber-toothed tiger, a big meat-eating cat that lived during the Ice Age and was about ten feet (three meters) long from nose to the tip of its tail. The Pleistocene period lasted from about two million years ago to about ten thousand years ago. Smilodons hunted in the wilds of what is now North America, where skeletons of this huge cat have been found at the La Brea tar pits near Los Angeles, California. These meat-eaters boasted upper canines that extended seven inches or more below the lower jaw.

☞ A 99.

There will be quite a difference in taste and appearance of the two carrots. The carrot that still has its leaves may show the first signs of rot because the sap will have continued to flow from the vegetable's root all the way to the leaves, with the result that it has deprived the part we eat of most of its nutrients and flavor. The leafy top will have started to wilt and the carrot itself will also show signs of rot. On the other hand, the carrot without the leaves will have retained virtually all its taste, vitamins, and "fresh" appearance.

☞ A 100.

A forerunner by several thousand years. The former is named after Alexander of Aphrodisias, who lived before Christ. The dark band that he discovered and brought to the attention of astronomers is the distinctly dark region between a bottom and top rainbow. While the bottom rainbow shows the usual colors of red, orange, yellow, green, blue, and violet, with red the most brilliant color, the second, higher rainbow, always shows the colors in reverse. (Sir Isaac Newton interpolated a seventh color to the rainbow, insisting that the color of indigo could be seen as a separate hue between blue and violet.) And what is referred to as "Alexander's dark band" is the space between the two rainbows, which is visibly darker than the sky that surrounds them. Also, being of a nonmaterial substance, the rainbow's spectral nature will always appear semicircular, no matter from what angle the observer views it.

☞ A 101.

Yes, England's John Ray (1627–1705) introduced a classification scheme for plants in 1667. He grouped them into monocots and dicots (flowering plants with one seed leaf and plants producing seeds with two cotyledons), which meant he based the classification on the number of their seed leaves. In 1704 he finished his major work *Historia plantarum* in three volumes, analyzing 18,600 plant species. This was the first definition of species as a branch of science. In his *Genera plantarum,* Linnaeus classified 18,000 species of plants in 1737, and he completed his *Species plantarum* in 1753, which clearly proves that Ray laid the groundwork for Linnaeus (1707–1778), although Ray did much of his work in natural history with his one-time pupil Francis Willughby (1634–1672). Nevertheless, Linnaeus must be recognized as the foremost botanist of his age and rightfully as the man after whom the world's system of classification of animals and plants is named.

☞ A 102.

Almost every night, but never in the Northern Hemisphere. If you were to go to Latin America, by the time you got to Argentina you could see these constellations. By the same token, Polaris and the Big Dipper are not visible anywhere in the Southern Hemisphere once you get as far south as Argentina. The reason for this is that our planet is permanently in an almost static orbit around an axis that varies little with respect to the location of stars and planets.

☞ A 103.

Henry David Thoreau (1817–1862). Besides his *Walden* (1854), he completed about thirty volumes based on his daily nature walks. The 1993 publication, newly titled *Faith in a Seed,* combines natural history writings and observations on the growth of communities. It took Bradley P. Dean (b. 1954), the editor of the *Thoreau Society Bulletin,* with the help of several Thoreau scholars, about ten years to sort out and collate the manuscript, which Thoreau had written, shortly before his death, on the backs of letters and broadsides and on hundreds of scattered pages to save paper.

☞ A 104.

Peppermint (*Mentha piperita*). Peppermint leaves produce a pungent, caffeine-free, and soothing tea. Mint is also used to flavor toothpaste and it has medicinal properties for gastric upsets. The last syllable of the herb's name derives from Pluto's beloved Mintho. When Pluto's jealous wife got wind of the affair she changed the poor nymph into a lowly plant.

☞ A 105.

Earth's seasons are comparatively stable because the moon's strong gravitational pull keeps our planet's tilt quite steady, stabilizing Earth and prompting its axis of rotation to remain at a constant 23° from the vertical.

☞ A 106.

In the United States it belongs to the genus *Phoradendron* and the family *Loranthaceae,* although the general public knows it better as mistletoe. In Europe the mistletoe is a semiparasitic green shrub (*Viscum album*), literally hanging around deciduous and evergreen trees and shrubs. It often cracks the bark of the host tree and literally sucks out its host's essential sap.

☞ A 107.

Both are correct. An obovate leaf is one shaped like an egg but with the narrowest point at its base. And an ovate leaf is also shaped like an egg but its broadest point is at the base.

☞ A 108.

The Greek name for testicles is *orchis,* and the common name of the plant is, of course, the orchid, while the botanical name is *Orchidaceae.*

☞ A 109.

Yes. A regular hen's egg has about 7,000 openings, or pores, in the shell. These openings permit the interior air molecules inside the egg to expand and escape as tiny bubbles when the egg is heated in hot water.

☞ A 110.

They are part of the tea plant, which is really a small tree or a shrub, after extensive pruning. The first and second congou are the lowest part of the tea plant, while the location of the plant's foliage above the congou is referred to as as the first and second souchong. More familiar are the next three higher parts of the plant, which are designated as pekoe, orange pekoe, and pekoe tips, the leaves at the very top.

☞ A 111.

In the botanical sense a jacket is the skin of a potato, especially when this vegetable is ready for boiling. A rhizobium is one of the beneficial nitrogen-fixing bacteria in the shape of nodules that are located alongside the roots of leguminous plants, such as peas, beans, some herbs, and shrubs, clovers, etc.

☞ A 112.

In fear of the Church's wrath, Copernicus did not allow the publication of *The Revolution of the Heavenly Spheres* (1543) until shortly before he died. In this great book he fully described and explained that the Earth revolved around the Sun, refuting the accepted stand of his time that our planet was the immovable center of the universe. Many of his vocal supporters were later unable to escape harsh penalties inflicted by Church and State.

☞ A 113.

This theory was scientifically accepted until 1992. But S.A. Zimov of the Pacific Institute for Geography in Vladivostok concluded that carbon dioxide is also released by microbes during icy conditions. Zimov measured 13.8 grams of carbon dioxide released per square meter from the frozen soil in northeastern Siberia early in 1993, while the Alaskan tundra released more carbon dioxide than it assimilated in the summer of 1992, according to Dr. Walter C. Oechel of San Diego State University. It is now believed that microbes discharge carbon dioxide throughout the year, resulting in the emissions of the greenhouse gas or effect. The latter is the consequence of carbon dioxide and other gases trapping heat in the atmosphere, similar to glass plates trapping heat in greenhouses.

☞ A 114.

This was the Danish astronomer Tycho Brahe (1546–1601). His highly accurate observations of the motions and positions of stars and planets proved to be the backbone of Kepler's three laws of planetary motion. In spite of Brahe's belief in a modified geocentric system, his detailed observations finally supplied Kepler with the vital evidence he needed to prove the validity of the course of our planetary system. At the age of twenty-seven, Brahe became one of the most celebrated European astronomers when he discovered a supernova in the constellation Cassiopeia. His observation and verification proved that the flare-up and eventual fading of this new star was tantamount to refuting previous astronomical suppositions that fixed stars were immutable and unchanging in their sphere.

☞ A 115.

They are very much alive and better known as weeping willows. This tree is actually an Asiatic plant belonging to the willow family, which includes over 300 species worldwide, such as the American pussy willow (which has the odd Latin name *Salix discolor*), the quaking aspen, and different species of cottonwood.

☞ A 116.

It's not even the nearest star to the North Pole today, let alone twelve thousand years from now. Since the pole has moved, its closest star is now Ursa Minor, and twelve thousand years hence Vega will be closest. This is due to precession of the equinoxes, which results from the gravitational pull of the sun and the moon upon protuberant matter about the Earth's equator. This means that the two points of the celestial and ecliptic equator—the equinoxes—gradually change their positions along the ecliptic—the circle of the celestial sphere that is the apparent path of the sun. They change every 25,800 years. For this reason ancient accounts of the stars (even for astrological purposes) no longer apply to today's sky. Particularly, monuments and observatories such as Egyptian pyramids and England's Stone-henge today point to different stars than they did at the time they were erected, as the opening remarks of this answer should make clear.

☞ A 117.

These plants are called passionflowers not because they serve as a love potion but because Catholic missionaries in the Western Hemisphere saw in the tendriled plant's convoluted winding stems a fancied semblance to the configurations that are manifested in Christ's Passion or crucifixion.

☞ A 118.

Epiphytes are plants that in their native state depend on other plants, mostly the tops of trees, for physical support. (They also fasten themselves to man-made structures.) The two best-known epiphytes are bromeliads and orchids, but some ferns also qualify. Epiphytes are native to tropical and subtropical regions, such as South and Central America, tropical Asia, the Caribbean, and Africa. The word *epiphyte* does not apply to a genus or plant family as such, but it derives from the Greek words *epi* (upon) and *phyton* (plant). Epiphytes do not depend for their livelihood on food provided by their hosts but just on their physical support; consequently they should be considered part of an ecological group with similar habitats and requirements, not as parasitic plants (such as mistletoes). For the most part epiphytes derive their sustenance of nutrients and moisture from air and rain. In temperate Western zones, the more well known epiphytes are found among mosses and lichens. It must be pointed out, however, that an epiphyte can also be a parasitic fungus growing on animal bodies.

☞ A 119.

No. Cicadas belong to the homopterous division of the order *Hemiptera* while locusts are members of the order called *Orthoptera,* which includes crickets and grasshoppers. Only male cicadas emit their characteristic shrill noise, which is believed to be a mating call.

☞ A 120.

Only an infinitesimal amount, since there is only about 0.03 to 0.05 percent of CO_2 in the air. Most of the carbon dioxide that is consumed is absorbed from the air by plants during the process of photosynthesis.

☞ A 121.

The flower and the dye are named after the German botanist Leonhart Fuchs (1501–1566), who was professor of medicine at the University of Tübingen for around thirty years. The first printed reference to the plant can be found in Father Charles Plumier's *Nova Plantarum Americanum Genera* (1703). The plant is generally believed to have been sighted for the first time in Santo Domingo. There are more than two thousand species, varieties, and cultivars of the genus *Fuchsia.* While some are dwarf shrubs, others can obtain a height of up to sixteen feet; still others are treelike with stems several inches in diameter. The dye is produced from the blossoms of some species of the genus *Fuchsia.*

☞ A 122.

Both use echolocation to find or avoid objects: dolphins in water, and bats in the air. Technically, echolocation can be defined as a method of locating objects by determining the time it takes for an echo to return and the direction from which it returns, as by sonar or radar. That dolphins could echolocate was established by Kenneth Norris and John Prescott as late as 1960 while scientists have known that bats could do it since the nineteenth century.

☞ A 123.

The animal is the shark. Scientists believe today that there is a vitally potent antibiotic in almost every cell of the shark's body, but it is an unknown class of antibiotics. Nevertheless, this compound happens to be closely related to cholesterol and is responsible for the shark's health. Biologists in the medical profession have determined, however, that the compound is quite effective against such microbes as bacteria, parasites, and fungi. Also, in Cuba and in the United States research has already established that in some cases shark cartilege injections have been successful in treating some patients with advanced states of cancer.

☞ A 124.

Most likely none of these men would have known about "naked smut." This disease is in the province of plant pathology. Naked smut, also called loose smut, is a disease of cereal grasses that is caused by the powdery smut fungi of the genus *Ustilago*.

☞ A 125.

In October 1967 Alf E. Porsild and Charles R. Arington, two American botanists, reported that they could grow arctic lupines from seeds that had been frozen for about ten thousand years—since just about the end of the last Ice Age.

☞ A 126.

It's more like a new wives' tale. If rice did cause the stomachs and crops of birds to explode, the avian population of Asia and the American bayous would be virtually nonexistent. Even though hungry pigeons, doves, cardinals, chickadees, finches, and other birds in urban areas consume rice, they prefer to still their appetite on different bird feed. Nevertheless, almost everywhere rice is considered a symbol of prosperity and fertility, and it is only lately that people have become so food-conscious that they consider it wasteful to spend precious human nourishment such as rice on birds since this food could be used more advantageously to feed the hungry.

☞ A 127.

In the scientific sense *macho* is an acronym standing for "massive compact halo objects." It represents the dark, invisible matter that surrounds galaxies. Astronomers believe that, since galaxies rotate so quickly, they would disintegrate if they weren't protected by this massive shield of dark matter. Some of the galaxies revolve around one another at such an astronomical speed, they need this surrounding dark mass of small dim stars or planets (*macho*), which creates more gravity, to prevent a cataclysmic collision of galaxies.

☞ A 128.

It was a famous African-American scientist named George Washington Carver (1864–1943). This botanist and chemist, whose parents were enslaved in Missouri, made about three hundred products from the peanut, which belongs to the *Leguminosae* family. Among the products he made from peanuts were coffee, flour, ink, cheese, milk, dyes, soap, and insulating board. Carver also made many products from the sweet potato, such as vinegar, molasses, rubber, and flour. Among his other inventions, he made synthetic marble out of wood shavings, and he was a pioneer in the field of plastics.

THE HUMAN BODY

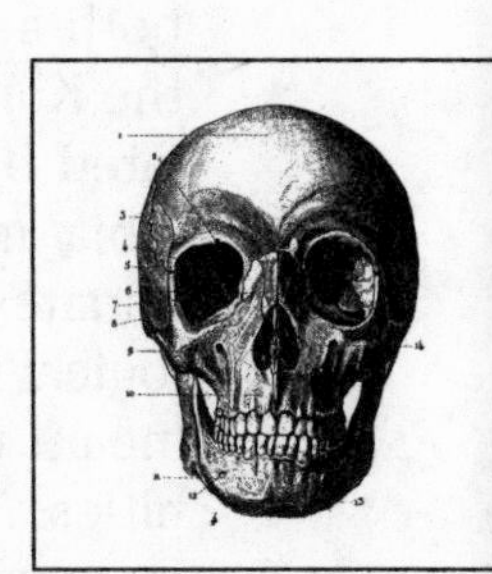

☞ A 1.

The most vital role vitamins play is to regulate and manage the chemical reactions in the body that protect cells and convert food into energy and healthy tissue. Vitamins are essential for the proper functioning of the body, and necessary for normal metabolism. Vitamin A, found in egg yolk and butter, helps the body grow and fights infections. B_1 or F is beneficial to nerve endings and assists in eliminating bodily wastes. B_2 prevents pellagra. Vitamin C, found in citrus fruits and several vegetables, is an anti-scorbutic (fights scurvy). Vitamin D prevents rickets, while K, which is found in some green vegetables, promotes normal blood coagulability. Of course, each of these vitamins also performs many other beneficial functions.

☞ A 2.

It is circumcision. Although circumcision is performed all over the world on millions of males for alleged health and religious reasons, the procedure is also performed on many Muslim females. This operation is described both as a cultural tradition and as a religious law, but the Koran does not demand it. Female babies are mutilated by having their clitoris removed as well as their labia and in extreme cases this irreligious practice culminates in the stitching up of the vagina. The physiological result prevents sexual pleasure and was initially meant to exert control over women in Muslim communities. The consequences of this medical practice on women has led to severe infections and occasional fatalities.

☞ A 3.

It is the formation by the liver of bile pigments; the coloring matter of bile, whose function it is to aid digestion.

☞ A 4.

Those without access to niacin (vitamin B_2: found in liver, blood, yeast, bran, legumes) and protein (meat, fish, cheese, eggs, vegetables) will fall victim to a sickness called pellagra. It is marked by dermatitis (inflammation of the skin) as well as gastrointestinal disorders and irreparable damage to the central nervous system. On the other hand, kwashiorkor is the result of food intake abnormally high in carbohydrates and abnormally low in protein. (Carbohydrates are stored as glycogen in muscle tissue and they can be found in candy bars, sweetened soft drinks, pasta, bread, potatoes, etc.) Kwashiorkor mostly affects children, and this extreme type of dietary deficiency is characterized by failure to grow and develop, by changes in the pigmentation of the skin and hair, by the fatty degeneration of the liver, and by anemia, edema, and total apathy.

☞ A 5.

These are the four kinds of fingerprint patterns. Although it is known that no two people have the same biological pattern of prints, one person may have just one kind or a mixture of fingerprint patterns.

☞ A 6.

Because they harden in less than five minutes, but can still be shaped by filing and carving for a quarter hour. The composition of dental amalgam consists of a minimum of 25 percent tin, at least 65 percent silver, and a maximum of 2 percent zinc and 6 percent copper, all of which is mechanically mixed with mercury.

☞ A 7.

Because mixoscopia is the medical term for the mental and psychological obsession with watching motion pictures (and live performances) of people having sexual intercourse.

☞ A 8.

Oddly enough, the human body has lost its ability to synthesize probably the best-known vitamin: Vitamin C. We can derive this necessary vitamin only from outside ourselves, by drinking plenty of orange juice, for example.

☞ A 9.

The nucleus of the human red blood cell has no name because it does not exist. In fact, our red blood cells are the only cells in the human body that do not contain a nucleus. True, they have a nucleus when they evolve in the bone marrow, but not long afterward the nucleus is expelled from the cell, and with no nucleus the cells cannot exist longer than four months.

☞ A 10.

It is the hepatitis C virus. To a large extent hepatitis C is transmitted through shared needles, infected blood, and in rarer cases through sexual intercourse. Even though patients infected by the hepatitis C virus do not die at a higher rate around the twentieth year of their infection, their death rate from liver disease (cancer and hepatitis) is higher than those suffering from hepatitis B. (Relatively few people infected with hepatitis B become chronic carriers of hepatitis.) Only about 25 percent of the patients of the C virus—the most widespread type of viral hepatitis—respond favorably to the only treatment available, the immune-system hormone alpha interferon.

☞ A 11.

It is *not* safe. During all sexual practices, condoms should be put on the male member *prior* to sex acts. As late as December 1992, it was clinically proven by Dr. Deborah Anderson, associate professor of obstetrics, gynecology, and reproductive biology at Harvard

Medical School, and her staff that the HIV (human immunodeficiency virus)-positive cells were present in pre-ejaculatory fluid. Known as CD4 receptors on the surface of healthy immune cells, these lymphocytic and macrophagic cells consume invading foreign bodies and are used, in conjunction with CD26 receptors, by the HIV virus to infect the immune cells. Consequently, CD4 and CD26 receptors are an essential part of the immune system and the main target for HIV. The number of CD4 receptor cells, for example, in infected HIV patients is liable to drop from the normal level of 1,200 cells per 1/1000 of a milliliter of blood to a dangerously low 200. However, in most AIDS patients, the CD4 count stabilizes at first around the 500 mark, which is not yet life-threatening. Oddly enough, there are long-term, healthy survivors with CD4 counts of less than 200. Immunologists believe that many white blood cells may replace the missing CD4s and reinforce the body's immunological system to fight off the terminal AIDS stage indefinitely. In terms of sexual practice, this means that there positively is a risk that during sexual foreplay this *pre*-ejaculatory secretion of cells will be released and the HIV causing AIDS can be passed on to the sex partner, even without the climactic ejaculation.

☞ A 12.

It attacks no organ and is anything but dangerous to the vast majority of humans. In fact, nicotinic acid ($C_6H_5O_2N$) can be found in protein foods like whole grains, nuts, soybeans, dried beans, eggplant, and some lean meat. Also known as vitamin B_3, it is just another name for the acid form of niacin. It is true that nicotine may have a chemical group in common with nicotinic acid but it is very far removed, and the latter can also be used as a medical drug to reduce cholesterol. If people are deprived for too long of nicotinic acid (and niacin), its deficiency can cause pellagra.

☞ A 13.

They are chromosomes, the microscopic bodies located in the nuclei of cells of almost every plant and animal. The fern has as many as 1,260, and the cells of the Australian ant have only one each. Humans have forty-six, and cabbage contain eighteen. As we all learned in junior high school, Gregor Johann Mendel's (1822–1884) garden pea has fourteen chromosomes per cell. Organisms that arise from the union of two sex cells have twice as many diploid as haploid chromosomes, i.e. the diploid number for men and women is 46, and the haploid number is 23. The word chromosome was first used in 1888 by Wilhelm von Waldeyer-Hartz, the German biologist, although they were originally identified as distinct cell phenomena by the Czech biologist Walther Flemming in 1873. Mendel did not break through with his discovery until 1900, although in all fairness it must be emphasized that he first published his findings in 1865, but they were temporarily lost or unrecognized. By 1865 he had already mentioned the existence of gametes as hereditary factors. These were later systematically connected with the genes carried by the chromosomes.

☞ A 14.

Fruit flies are especially good because their simple chromosomes are perfect for observing aspects of heredity—and because they breed fast. And fruit flies, including their living embryos, are the most complex organisms ever to be frozen cryogenically in labs and revived.

☞ A 15.

The thymus is a central lymphoid organ in the neck or upper thorax. It obtains its relative maximum weight at birth, almost half a percent of the baby's total body weight, and expands in size and weight until puberty is reached (about 37 grams), but thereafter it is reduced in size (it involutes) and in the seventh decade of a person's life only weighs about six grams. It plays a role in the production of lymphocytes (white blood cells that are part of the body's immunological defense against infections), although some scientists believe it may also be partially responsible for the biological course of growth and sexual development. Otherwise it is still an organ puzzling scientists. The thyroid, on the other hand, has a more definite objective, well known to the medical profession. This vascularized endocrine gland is attached to the thyroid cartilage of the larynx below the Adam's apple and produces, stores, and secretes the thyroid hormones, a function controlled by the pituitary. Weighing about one ounce, it determines the level of metabolism in the blood and the rest of the body, regulating human growth, physical development, and maturation; it also counteracts, for example, the presence of too much calcium in the blood.

☞ A 16.

On the contrary, it has increased astronomically. Just a few years down the road, by the year 2,000, between 1 and 2 percent of the U.S. population will develop skin cancer, as compared to only 1/15 of 1 percent about sixty-five years ago. Melanoma will strike the younger generation between the ages of thirty and fifty most of all, since the young will not heed the warnings about increased exposure to the sun's radiation. An equal contributor to melanoma is the increasing depletion of the ozone layer that protects the planet we live on from the sun's radiation. By the year 2,000 scientists are projecting a 7 percent loss of the ozone layer, with the result that the *mortality* rate of skin cancer will increase significantly, maybe by about 10 percent. Although about 7,000 Americans die annually from malignant melanoma in the 1990s, an additional 2,300 are killed each year by the squamous-cell carcinomas that result when the skin's upper layers are exposed too long to the sun's rays.

☞ A 17.

As a result of a viral infection, the attacked cells frequently will produce special proteins and transfer them to neighboring cells. The purpose of this diffusion of proteins is to temporarily create a resistance to the invading virus. Without this protein the virus would succeed in reproducing itself inside the neighboring cell. These defense mechanisms of the human body are called interferons. In order to produce these healing proteins, genes must first be activated to manufacture them, and this is aided by the relatively new recombinant alpha interferon. Treatment with alpha interferon has been successful in cases involving certain types of hepatitis and leukemia.

☞ A 18.

It's not a matter of falling headfirst on tightly packed snow, although that can cause more damage to a skier than a headache. A large number of the headaches skiers get are caused by altitude. Many people cannot tolerate the air pressure and the reduced amount of oxygen available at altitudes above 7,000 feet. Although many skiers gradually become accustomed to the altitude, others fight altitude-induced headaches with cortisone and diamox. Nevertheless, acute mountain sickness can prove fatal, unless the sufferer immediately retreats to a considerably lower elevation.

☞ A 19.

Allergic rhinitis is more commonly known as hay fever, and the botanical bearers are pollen grains. Unfortunately, they also aim for other targets that are sticky and moist, such as mankind's bronchial tubes, eyes, and the mucuous membranes of the nose. Although the term *hay fever* was coined in 1829 by the *Medical Gazette,* the malady is rarely responsible for a fever, and hay has nothing to do with it. No vaccine has yet been manufactured to cure this allergy.

☞ A 20.

The microscopic structure of cork Robert Hooke demonstrated in his book clearly displayed the boundaries of cell walls. He referred to them by the Latin name *cellulae,* and from this word the biological term *cell* derives. In the same book he also showed the cells of a stinging nettle.

☞ A 21.

The command center of genes for giving hereditary instructions is temporarily held in abeyance, in order to give the new reproductive cell a chance to "teach" its progeny how to manufacture and combine substances (e.g., proteins) required for future chemical compositions in the body. The information must be passed on to the cells prior to their actual division, or the new cells would not "know" how to program subsequent generations.

☞ A 22.

The scientist was Robert Koch (1843–1910). In 1876, this German bacteriologist isolated a rod-shaped bacteria from the blood of anthrax-diseased cattle and injected this pure culture into a healthy animal. The same kind of bacteria were isolated when symptoms of anthrax appeared in the animal that had been healthy. This process resulting in a second isolation of the same sort of organism, was the proof of Koch's postulates. The bacterium that causes anthrax is transmissable to man and the resulting disease is characterized by lesions in the lungs or external ulcerating nodules. After devising the means to culture these bacteria outside the body, Koch formulated rules to determine whether a bacterium was the cause of a disease. As a consequence he won the Nobel prize in medicine in 1905.

☞ A 23.

About 100 trillion (called 100 billion in some European countries).

☞ A 24.

The sand is not supposed to glue the lids to the bottom of our eyes so we can go to sleep. What we call "sand" is nothing but dried mucus. Sleepiness, like any other internal or external irritation (e.g., disease, smoke, or cut onions) causes the secretion of mucus. With the increase of daily or nocturnal irritation to the eye, the mucus tends to expand and gathers in the corners of the eyes. When we are asleep, however, the closed lids retain the moisture of the eyes, although the moisture's mucus may gather in the corners, dry out, and cause us to wake up with "sand in our eyes."

☞ A 25.

When a spinal cord has been severely damaged, the cervical nerve cells inside it begin to die and, almost like decaying flesh, they discharge toxins. These toxins are renegade oxygen molecules that attack nearby undamaged cells, and it is these molecules that are called free radicals. They use a flood of biochemicals to destroy even more neighboring nerve cells. These toxic molecules are reactive, unstable compounds that carry an extra electron, which allows the free radicals to react with other molecules damaging the cells' protective membranes and altering the genetic information in the cells' DNA. This creates more free radicals. Some of the damage done by free radicals can be prevented by vitamins C and E, beta-carotene, and such trace minerals as selenium, zinc, and manganese as well as by eating plenty of vegetables and fruits.

☞ A 26.

The follicles produce a strand of hair that emerges through the skin pore. The follicle could not show off its growth without the pore.

☞ A 27.

Freud was proudest of his 1900 opus *The Interpretation of Dreams.* On the other hand, many psychiatrists consider *Moses and Monotheism,* written at the age of eighty-three and published the year he died (1939), to be Freud's most significant paper.

☞ A 28.

Hydrochloric acid actually kills some of the microorganisms in our stomach. During this process it creates an enzyme called pepsin (secreted by glands in the stomach's membrane), which is largely responsible for disintegrating and digesting most proteins we have ingested. Similar processes go on when enzymes are manufactured in the intestine wall (as well as in the pancreas and liver), and their primary function is to dissolve our food's proteins, fats, nucleic acids, and carbohydrates so that they are diluted sufficiently to be ingested into the small intestine. In the last stage of digestion the large intestine removes nutrient liquids from the resultant mass, and the remaining pulp is expelled through the anus as waste material.

☞ A 29.

Simply put, it can take two or three decades of exposure to certain foods before the terminal immune response is sufficiently provoked to cause an allergic attack. People can break out in hives or suffer asthma attacks, but in most cases those subject to these attacks frequently have had other less pronounced allergies. Frequent consumption of certain foods tend to end up causing allergies. That's why rice allergies are most widespread in Japan and codfish allergies abound in Scandinavian countries. But inherited biochemical transferences can also add to these ailments, as can the handling of household pets. When the genetic threshold of tolerance to food or other allergens has been crossed, allergies that had not been noticeable before will suddenly emerge. However, it is extremely rare for a first incident of food allergy to develop in middle age.

☞ A 30.

It is a single molecule of human DNA. This feat of compression is performed to a large extent by histones, five proteins that squeeze and cram the DNA down to size.

☞ A 31.

Dopamine controls the thought process that determines how we function physically, even the slightest movement of our fingers, the nod of the head, and the curling of toes. Chemical substitutes, such as L-dopa, plus the transfer of fetal tissue from healthy brain cells to the diseased brain part in an afflicted person, may eventually be instrumental in curing patients suffering from Alzheimer's or Parkinson's or other degenerative diseases that limit the control of bodily movements.

☞ A 32.

Phacoemulsification deploys ultrasound to remove cataracts (which cause clouding of the eye's natural lens) in tiny pieces. With even newer medical advances, phacoemulsification has led to small-incision-cataract surgery, which in turn led to the development of a still newer procedure known as No-Stitch surgery. In this latest technique—the tunnel incision—a microscopic flap is cut into the sclera (the firm white membrane forming the outer coat of the eye) and the surgery is performed through this flap. Immediately after the removal of the cataract, the eye's internal tension locks the incision up airtight, so that in most cases no stitches are required. This allows for a rapid recovery of vision, and most patients can return to an active life-style within a day or so after surgery.

☞ A 33.

Dr. Rosebury, in his 1969 classic *Life on Man,* referred to normal microbes living on and inside the human body. Microorganisms live in untold numbers in and on each person, healthy or sick. Billions of these organisms are not only benign but necessary to our well-being. For instance, saliva from a healthy person can have as many as a billion bacteria per cubic centimeter, and tiny mites are known to live harmlessly at the base of our eyelashes throughout our lives. Dr. Rosebury points out that as many a five million bacteria can live on each square centimeter of human skin. They serve as scavengers and help to ward off disease.

☞ A 34.

Sigmund Freud always credited his friend, Dr. Josef Breuer (1842–1925), as cofounder of psychoanalysis. They wrote the first paper on hysteria in 1895, a paper that contained the first published references to free association and transference. A short time later, Breuer began to question Freud's emphasis on children's sexuality and his association with the Berlin ear, nose, and throat specialist, Dr. Wilhelm Fliess. Fliess had influenced Freud's views on children's sexuality and later frightened him with superstitious speculation that the Viennese doctor would die in February 1918. Equally important, Freud's intimate relationship with Dr. Fliess ended in animosity when Freud bitterly disagreed with Fliess's belief that the nose was the body's controlling and dominant organ. This breakup, based on intellectual disagreements, closely mirrors Freud's falling out with his celebrated friend, the psychiatrist Carl Gustav Jung (1875–1961), who had largely embraced Hitler's National Socialist racial policies.

☞ A 35.

It is the Spanish and French acronym for AIDS. In Spanish SIDA stands for "Sindrome de immunodeficiencia adquirida," and in French for "Syndrome immunodéficitaire acquis."

☞ A 36.

All these terms describe techniques for tracking neurological activity. PET charts brain activity by tracking the blood flow within a radius of five inches. MRI shows detailed images of the brain. SPECT tracks blood flow and brain activity. SQUID picks up the brain's magnetic field and brain activity.

☞ A 37.

Patients taking aspirin or other similar painkillers to combat the flu or a cold, for instance, soon realize that their fevers are lowered. And that's the problem. A moderate fever is an important mechanism in the body's defense against unwelcome invaders. The primary purpose of a high temperature is to activate and energize the body's white blood cells, enabling them to rush to the infected area and destroy the malignant microbes there. Moreover, the heat generated by a fever helps to inhibit the growth of intruding bacteria that become sluggish and finally disintegrate. Body temperature rises in most animals to burn off pathogens, and even cold-blooded animals like iguanas and lizards lie in the sun to make themselves feverish when not feeling well.

☞ A 38.

A few. For example, exposure to sunlight permits the creation of vitamin D in our skin. Three other vitamins are manufactured inside the human gut by resident bacteria: vitamin K, pantothenic acid, and biotin. But biotin can also be ingested by eating egg yolks and liver; vitamin K by consuming leafy vegetables, dairy products, liver, and cereals; and pantothenic acid is an oily acid, a member of the vitamin B complex, and it may also be found in meats and fish.

☞ A 39.

In your body—in the brain to be exact. The sea horse–shaped structure consolidates recently acquired information and turns short-term memory into a storage chamber for long-term retrieval. H. major extends the whole length of the cornu (a hornlike process) in the cerebrum, and H. minor projects backward into the posterior lobe of the cerebrum, which is the chief portion of the brain and fills the upper cavity of the skull.

☞ A 40.

False. Cataracts can be removed at any stage of development, although it is advisable that they be eliminated before the patient feels that they interfere with his or her vision.

☞ A 41.

Nobody knows. Logically one would think that with the production of testicular hormones diminishing, the prostate would shrink. But the exact opposite is oddly the case. The enlargement of the prostate, which can be cancerous, may have something to do with sex hormones since eunuchs are virtually never afflicted with an enlarged prostate. What the medical profession does know is that once the male hormone testosterone converts into dihydrotestosterone, this latter substance is responsible for the prostate's enlargement. Nobody, however, knows what causes this substance suddenly to expand the prostate or why it should do so after men reach their fiftieth—sometimes their fortieth—year of life.

☞ A 42.

Sigmund Freud's studies on aphasia (the loss of ability to articulate words) are still highly regarded by neurologists, as is his research on the cerebral paralyses of children. Freud was in his early thirties when he did his work in these subjects.

☞ A 43.

Medical students may know that most people have two deep horizontal lines in the palms of their hands and that they are called simian creases. In just a few of us these two creases join up and blend together to form one straight line crossing the palm. They as well as the other lines in our palms—such a favorite with fortunetellers—are formed quite late in the development of the fetus—as it begins to extend and flex its hands in the uterus. In contrast, the fine fingerprints, particularly the dermal ridges (the "engraved" whorls on the outer layer of skin), are already formed in the fifth month of pregnancy.

☞ A 44.

It is no secret that a fat-free diet can prevent complete blockage of artery walls. But why do the linings clog up in the first place? When too much cholesterol is taken into the body, the natural oxidation that takes place results in a rancid accumulation of fatty substances that interfere with blood circulation. The oxidized cholesterol attracts other cells, and together they form artery-blocking plaque. The more plaque, the greater the damage which can cause a heart attack. Drugs may help (with possible side-effects) to prevent oxidation of LDL cholesterol as do some vitamins, notably antioxidant vitamin C, vitamin E, and beta carotene.

☞ A 45.

The only part of the human body that cannot develop cancer is the lens of the eye.

☞ A 46.

Lasers cannot be used to remove cataracts. They are used, however, to improve vision *after* the cataracts have been surgically removed. Sometimes membranes can become hazy as a result of the cataract removal. In this case lasers are useful in opening the membranes. For example, once a cataract extraction has been performed, the neodymium-YAG laser can create an opening in the posterior capsule remaining in the eye.

☞ A 47.

It is not the bones themselves that crack when people crack their knuckles. Knuckles and other human joints are blessed with a certain amount of elasticity. When a finger joint is suddenly yanked and sharply repositioned, a vacuum is created between the bones. The consequence is that fluids displaced by the artificially created vacuum rush back into that empty gap, and it is this sudden surge of fluids into the vacuum that causes the popping sound.

☞ A 48.

It is the instrument that measures arterial blood pressure.

☞ A 49.

The four basic tastes that can be distinguished by the receptors of the taste buds on a human tongue.

☞ A 50.

Pellagra is caused by a deficiency of niacin (one of the B vitamins) and is characterized by alimentary-tract, skin, and nervous-system dysfunctions. This virulent disease can terminate in maniacal behavior.

☞ A 51.

There are over 100 million light receptor cells on a person's retina. The color blue is most difficult to see as a person grows older.

☞ A 52.

Salicin and its white, crystalline solid, known as salicylic acid, are found in the roots and bark of the willow and are used to make aspirin.

☞ A 53.

None of these. It is chlamydia, which is a common pathogenic microorganism transmitting bacteria or viruses. The serotypes of *Chlamydia trachomatis* cause nongonococcal urethritis in men and cervicitis in women. There are three million infected with this disease in the United States per year as compared to two million gonococcal infections.

☞ A 54.

Approximately 200 million sperm enter a woman's vagina during an average ejaculation. Of these, only about fifty reach the woman's egg. The others are destroyed by the vagina's acidity or lose their way to the Fallopian tubes or attach themselves by the millions to round bodies they mistake for an egg. The head of the sperm that penetrates the egg splits wide open to release its genetic material, which combines with the hereditary message inside the ovum to begin the creation of an embryo.

☞ A 55.

Because Asian immigrants experienced difficulty with the English language at first, they concentrated on nonverbal subjects, such as science, accounting, mathematics, computer technology, and music. In the late 1980s more than 25 percent of all students at New York's Juilliard School of Music, for example, were young Asians. Their aspirations and success may well stem from the tightly knit Asian family unit, self-discipline, and parental love.

☞ A 56.

The uterus during the final stages of pregnancy.

☞ A 57.

The U.S. military authorities, who followed Spanish rule, were responsible for improvement in the education, health, and economic sectors in Cuba from 1899–1902. By the time Fidel Castro (b. 1927) came to power in 1959, yellow fever had been eliminated in Cuba for almost sixty years.

☞ A 58.

The semen protecting the sperm (or spermatozoa) does not flow at a uniform speed to the female egg (or ovum or gamete). Soon after entering the vagina, the semen comes to a near standstill and coagulates into a uniform mass to fight its way through the acidity of the vagina, which recognizes the semen (and 200 million sperm) as an unwelcome foreign body.

☞ A 59.

Freud (b. 1856) did not fight in World War I. He was much too old to serve. What gave him cause to superstitiously believe he would die was the persuasive power of the Berlin physician Wilhelm Fliess, who, quoting his own lunatic theory of biological rhythms, convinced Freud that he was to meet death in February 1918. Freud died in 1939.

☞ A 60.

Arteries lead the blood from the heart to the body tissues. The conducting arteries go directly from the heart while the distributing arteries channel the blood to all the principal organs. The delicate arterioles deliver the blood to the capillaries. The veins return the blood to the heart. The walls of arteries must be thick because the blood inside is under constant pressure. The walls of capillaries consist of a single layer of cells, enabling arteries to connect with veins and allowing oxygen molecules and plasma to pass through into the body fluid, and so that white blood cells can move to and from the bloodstream. The walls of veins are thin; they do not have to withstand the high pressures that the arteries must tolerate. In fact, vein walls collapse if the blood is drained from the veins.

☞ A 61.

None! The common cold is a viral disease, and antibiotics do not attack viruses. Antibiotics are chemical substances produced by various microorganisms and fungi that can inhibit the growth of or destroy another microorganism or bacteria in dilute solutions. Vaccination is the most common means of preventing viral infections such as polio, typhus, smallpox, influenza, etc.

☞ A 62.

There were just over 50,000 medical doctors in Germany before 1933; 80 percent of them were *not* Jewish. There were about 10,000 Jewish doctors and

42,000 non-Jewish doctors treating a nation of approximately 66 million citizens.

☞ A 63.

The hormone is progesterone. It helps keep the uterus in condition to accept implantation of the embryo. Biochemist Dr. Karl Slotta (1895–1987) discovered progesterone in 1935 at the Chemical Institute in Breslau, Germany, just before he and his family fled the Nazis for Brazil. Slotta also found that progesterone sends chemical signals to the brain to prevent ovulation. His research led to the development of birth control pills.

☞ A 64.

Lactic acid. Human energy sources are derived from the oxidation of sugars and fats. Combined with the anaerobic breakdown of carbohydrates they produce lactic acid in muscle contraction. This also applies to the animal kingdom. Just as water and carbon dioxide are released as waste products, so, too, is lactic acid after it has served its purpose. This latter elimination occurs when the body's glycogen, stored in the liver, is used to produce creatine phosphate, a substance stored in muscles to create energy.

☞ A 65.

There are many reasons. For instance, the sugar level of a diabetic person may actually close retinal blood vessels, resulting in the reduction of the retina's vital oxygen and blood supply. Or it can cause weakened blood vessels to bleed and leak serum into the retina, which will cause the retina to swell (retinal edema) and to collect fats from the blood's serum. Another side effect of diabetes is that abnormal blood vessels grow on the surface of the retina and as a result scar tissue develops on the retina, which can become detached. Another consequence of diabetes is that the eyes may hemmorrhage, a problem known as proliferative retinopathy.

☞ A 66.

There are only four: nailbeds, eardrums, the bulbous end of the penis, and part of the lips.

☞ A 67.

The wormlike larvae of flies we refer to as maggots are mostly the eggs (larvae) that blowflies deposit on corpses exposed to the air and remain unattended; they sate their hunger by consuming decaying parts of open flesh wounds. Maggots rarely infest the human anatomy unless they are in spoiled meat we have ingested. It is true that roundworms, hookworms, and tapeworms can live inside people throughout life, but as a rule they die shortly after the human being expires.

☞ A 68.

Glaucoma has no accompanying symptoms—at least not until it has damaged the patient's peripheral vision. Glaucoma usually develops silently and painlessly in people who are over forty years old.

☞ A 69.

The sprite, guarding some hidden treasure underground, and the aphorism are not genomes but gnomes. The genome is a complete set of 46 chromosomes with genes that are contained and found in the nucleus of all 100 trillion human cells, except in the red blood cells, which have no nucleus. There are an estimated three billion chemical code letters, or nucleotides, in the genome that form the building blocks and strands of human DNA. In fact, there is a Human Genome Project and a National Center for Human Genome Research whose mission is to determine the precise sequence of nucleotides. The purpose of these organizations is to map the genome. This entails locating and identifying ca. 100,000 genes and the sequence of their constituent chemical parts. This will assist microbiologists and molecular geneticists to replicate the precise order of nucleotides (or gene subparts) that make up our genetic code. The end result of this research may help scientists discover the cause and cure for humankind's most deadly diseases.

☞ A 70.

The polymerase chain reaction is used to make millions of copies of DNA (deoxyribonucleic acid, the substance of the genes) fragments by a process called gene amplification. This method is applied to a strand of hair, or small samples of blood, or semen found at the scene of a crime. Although the lab test does not pin down the exact identification of the DNA source, polymerase chain reaction can determine from which particular suspect the sample may have originated. Other suspects can be eliminated if they show no precise kinship to the DNA sample. Gene amplification is now considered a more discriminatory detective tool than tracing criminals through blood samples alone. Scientists can use polymerase chain reaction testing to determine biological factors in DNA molecules dating back millions of years.

☞ A 71.

Recent medical research has shown that adult-onset diabetes is not so much caused by defective insulin production by the pancreas, which regulates the metabolism of sugar and fat, as by a change in the utilization of the insulin by muscle cells. *Diabetes mellitus* is characterized by a disorder of carbohydrate metabolism, an excess of sugar in the blood and urine, which results in an abnormal appetite and a gradual loss of weight.

☞ A 72.

To answer the last question first, microbiologists established in 1992 that it would be self-defeating to replenish the supply with iron supplements. The reason for the iron-level drop of up to 90 percent of the normal supply is that the body's immune system labors hard to keep this nutrient metal away from bacteria, parasites, and fungi because these microbial enemies need iron to survive and divide. When the body is infected, a chain of enzymatic reactions immediately goes to work and carries most of the bloodstream's iron to the liver, where the bacteria cannot reach it. Depriving the alien bodies of needed iron and burning them up with a high fever is an immune-system double whammy.

☞ A 73.

ATP is the body's energy deposit. Its source is the mitochondria, which send ATP throughout the body wherever energy is required. These minute cellular bodies have their own genetic code and DNA, existing apart from the nucleus and responding to our bodies' physical needs and chemical demands. If the mitochondria work insufficiently in using oxygen to stoke the fires of chemical energy, the electrons that they normally release will not be able to change into ATP, and undesirable levels of lactic acid will consequently accumulate. The cell nucleus produces about two hundred genes to fabricate proteins, and if a specific one of these genes goes awry, the function of the mitochondria is upset, leading to the first clinical symptoms of Huntington's disease. This ailment, in which brain areas controlling normal physical activities are slowly destroyed, is characterized by mental deterioration and uncontrollable body movements. So far there is no cure.

☞ A 74.

Frequently used in hospital operation theaters, a capnometer is a device that gauges, quantifies, and evaluates the carbon dioxide exhaled by a patient during an operation. This enables the anesthetist to administer the correct amount of anesthesia and oxygen to the patient. New York State, for one, requires the use of this equipment in lengthy, complex operations.

☞ A 75.

It's the man's prostate. Its main function is to secrete nourishment into the milky fluid that carries the sperm (produced in the testicles). The male's secretions lower the acidity of the vaginal canal.

☞ A 76.

Many more people die from insect bites than from spider bites. The reason for this is that spider fangs are quite fragile and that people are highly sensitive to the much more piercing stings of bees, hornets, ants, wasps, and mosquitoes. And these insects are much more widespread in most countries than are spiders. The bite of insects such as mosquitoes can be fatal since some carry germs that can cause diseases such as malaria. The most poisonous spider is believed to be the black widow (*Lactrodectus mactans*), but it does not carry much poison (certainly less than a snake), and only about one percent of those with untreated black widow bites suffer fatalities (against twenty percent for the rattler). Still, before a bite is treated, black-widow venom can cause excruciating abdominal pain, convulsions, paralysis, and shock.

☞ A 77.

By mother's milk. All human mothers, regardless of breast size, have between fifteen and twenty lobes of milk-secreting glands, and the baby's sucking stimulates nerve impulses all the way to the mother's anterior pituitary gland, where prolactin is transmitted once more by the blood to the breast, this time to sustain the secretion of mother's milk.

☞ A 78.

A 1992 epidemiological study in Finland by Dr. Jukka T. Salonen has proven that higher blood levels of iron are riskier as far as heart attacks are concerned. One reason that premenopausal women suffer fewer heart attacks than men may be that a great deal of iron is flushed out during menstrual periods. It has been proven that for each increase of one percent in the blood's ferritin (the protein that binds iron in the blood) the danger of having a heart attack rises fourfold among men. But it has also been proven that some fish oils and aspirin, even though the latter may cause small amounts of bleeding, can help to protect the body against heart attacks. How much the female hormone estrogen contributes toward the low incidence of heart attacks in premenopausal women has not yet been determined. Scientists now assume that there may be an increase in heart attacks not only if the LDL cholesterol level proves high, but also if the level of the iron content in the blood is on the increase.

☞ A 79.

Simply put, they are water-soluble proteins found in cell nuclei. But what once was dismissed as their sole ability to condense and pack our DNA genetic material, today is regarded by the scientific community with much greater respect. Histones are actually capable of turning genes on and off along the DNA molecule, and this enables each cell to perform its life-sustaining activity. For example, histones can "order" the secretion of metabolic hormones if that task must be performed by a thyroid gland cell, or they can "demand" that bile salts be removed or stored to suit the needs of the gall bladder.

☞ A 80.

No, it's not something to please us. On the contrary: it's a disease that's been around since 2000 B.C., about 200 million people are affected by it, and it is caused by parasitic worms that live their entire lives mating in a blood vessel. The disease is schistosomiasis, and it is prevalent throughout Africa, the Arab countries, and much of Latin America. The great mystery of this disease is how the worms manage to escape all the body's defenses. They actually live and produce thousands of eggs daily inside the body's white blood cells—in masses called granulomas—where they are safeguarded and the immune system cannot reach them. One species of these worms, *Schistosoma mannsoni,* enlists the immune system hormone that in the normal course of events releases white blood cells in response to an inflammatory infection. This hormone (tumor necrosis factor) is used by the worms to locate their permanent home: the blood vessels around the intestines, liver, and bladder, where they can live to sexually stimulate each other continuously in order to lay eggs and procreate. Schistosomiasis is fatal, but one ray of hope is that the drug praziquantel can temporarily relieve patients of the disease.

☞ A 81.

The enamel of your teeth is the hardest substance in your body; 96 percent of it is made up of a concentration of mineral salts. It has no nerve supply, although the dentin, which is the body of the tooth and is made up of 70 percent of calcium and phosphorus, does supply it with tiny amounts of nourishment. Rods (prisms) of apatite within the enamel help to prevent corrosion and fragmentation during the chewing process. Besides the 96 percent mineral salts, the remaining 4 percent is made up of organic matter and water.

☞ A 82.

On the contrary. The body's response is to make matters even worse. The damaged, inflamed portion tends to cut off blood supply that is vital to the surrounding area, resulting in damage to additional nerve cells. Recently, in 1990, it was discovered that this flow of tragic consequences could be partially halted by injecting enormous doses of a steroid, methylprednisolone, within eight hours after an injury. Yale University's Dr. Michael Bracken and his staff found that this drug, by binding itself to the oxygen free radicals, prevented the radicals from attacking nearby tissue, thus decreasing secondary nerve damage by as much as a half.

☞ A 83.

Viruses have no common ancestors as far as is known, and therefore they are not phylogenetically (evolutionarily) related to any known viable substance.

☞ A 84.

The Goethe Prize, which Freud received in 1930. However, it was not awarded for his scientific findings but for his literary output. In spite of being one of the great geniuses and originators of what he considered to be a science—psychoanalysis—Freud was never awarded a Nobel Prize.

☞ A 85.

The statement is *not* true. After somatic death, human organs inside the deceased's body cease functioning, or die, at different times. For example, some neurons in the brain do not die for about five minutes after the final heartbeat and kidney cells survive a person's expiring breath by about thirty minutes. In fact, the muscle cells (though not the entire muscle) of the human heart survive a patient's death by another fifteen minutes, and liver cells hold the record at about three-quarters of an hour.

☞ A 86.

He found that old cells carrying these bacteria became attenuated (weakened) and did not cause cholera in the healthy chicken. Even virulent bacteria (after abovementioned primary inoculation) did not affect the chicken. Immunity had been induced by injection. The labors of Pasteur and Koch in the control of bacterial diseases were instrumental in tripling the life expectancy of humans between 1876 and 1926.

☞ A 87.

The disease is AIDS, and the human immunodeficiency virus (HIV) is the culprit. The virus somehow incapacitates human antibodies, the immune-system cells that regulate the body's defense. HIV frequently combines with the genetic material of the cells it has infiltrated, causing the patient to test positive for HIV without having AIDS. This leads some scientists now to believe that HIV itself may not be the causal factor of AIDS.

☞ A 88.

If gum and tooth diseases remain untreated, the body's resistance to infection will be lowered, with the result that in chronic situations life can be threatened. Periodontal disease starts when harmful bacteria and their by-products damage gum and bone tissues, which in turn generate sulfur compounds that cause bad breath and can infect organs throughout the body.

☞ A 89.

It's far from an old wives' tale. The sad fact is that people who suffer from bulimia, for example, the eating disorder in which patients alternately overeat and then purge themselves by vomiting, will bring up an uncommonly large amount of hydrochloric acid from the stomach. This acid tends to disintegrate and corrode the enamel and dentine of the teeth, especially the back teeth and the backs of the upper front teeth. Moreover, it can cause other periodontal problems, such as dulling the edges of the teeth and corroding fillings. To prevent greater damage to their teeth, bulimics should rinse their mouth with water and baking soda after each vomiting in order to neutralize the hydrochloric acid. Fluoride medications may also have a palliative effect.

☞ A 90.

In simple terms our hands turn blue from cold because the brain does not want the blood to freeze when external temperatures get too low. So the tiny blood vessels in your hands—the capillaries—contract, and as a consequence less blood is able to run through them to get chilled; the result is that the healthy color of the hands changes to something resembling gray, and even blue. Also, some of the blood rushes to the brain to keep it warm, because once the brain freezes and shuts down, all natural functions of the body will cease and death will occur. Before that happens, however, if you are exposed too long to the bitter cold, the extremities, such as your fingers and toes, even your hands and feet, may suffer irrevocable frostbite and require amputation. By the same token, if you are too hot, the blood vessels in your skin open wide to expose your blood to the cooler air outside your body, and your face turns redder and redder as a consequence. The perspiration emitted through your skin is also meant to lessen the outside heat on your skin.

☞ A 91.

Yes, it can. Blood plasma is used in transfusions. Plasma is almost colorless, because the red and white blood cells have been removed from its fluid. In fact, plasma makes up just over half of the content of human blood. The other ingredients are salts, protein, water, hormones, waste products, and nutrients. Unlike whole blood, plasma can be stored in a dry state for long periods of time without deteriorating.

☞ A 92.

Not too well. Especially if you have to pay a visit to your doctor. Because splanchnology is the branch of medical study dealing with the structure, functions, and diseases of the viscera—ailments relating to your skull, thorax, abdomen, intestines, lungs, liver, kidneys, etc.

☞ A 93.

The drug is a stimulant called amphetamine, also known by the trade name Benzedrine. It has been subject to widespread abuse. In spite of side effects such as hypertension, heart palpitations, tremors, insomnia, aggressiveness, anxiety, and paranoia, this stimulating drug is frequently "snorted" (using a derivative called methamphetamine) or "mainlined" (injected in dilute form intravenously). Frequently it is mixed in "binges" with large doses of depressants, giving the illegal user "uppers" and "downers." And it has been combined with cocaine and heroin, the latter application being referred to as a "speedball." Athletes have used it illegally to increase the quality of their performance.

☞ A 94.

There is no reason to celebrate. A more appropriate and popular word for this long-winded medical term is valley fever. It is a harrowing disease caused by the wind-blown microscopic spores of a fungus, *Coccidioides immitis.* The most afflicted parts of the U.S. are the dry areas of the West and Southwest, but it is also prevalent in Mexico and Central and South America. Thousands of victims are infected annually by the tiny spores, although only a small percentage of cases prove fatal. High fever, headaches, sores, and swollen feet, arthritic pains, and a mind-numbing exhaustion can be symptomatic of this affliction. The fungus often multiplies during protracted dry spells, when the infinitesimal spores can travel hundreds of miles on the top of anything as minuscule as dust particles. Some patients are helped by the removal of spinal fluid and its reinjection with the unsavory medication amphotericin B while others find temporary relief with the expensive antifungal drug fluconazole.

☞ A 95.

There is no blood test yet available to detect gonorrhea. Diagnosis has to be made by the isolation of the gonococcus in culture from the actual infected site. Unfortunately, so far no nonculture tests which identify antigens (substances introduced into the body to stimulate the production of antibodies), are available for diagnosing gonococcal infections in women, but a form of nonculture testing can be used to detect gonorrhea in men. Although gonorrhea is the most widespread communicable disease in the U.S. (after chlamydia), Europe, and developing countries primarily among the young, a noticeable decline in this disease has been reported in Europe and the United States.

☞ A 96.

Undoubtedly, the disease killing more people than any other since the early 1800s (the chief cause of death in the United States until 1909), and again on the rise, is tuberculosis. In those two hundred years it killed at least 1,000,000,000 people worldwide. Can TB be called the most malignant sickness ever? Far from it. Patients afflicted with heart disease and cancer have a higher mortality rate percentage-wise than those suffering from tuberculosis. The most widespread form is pulmonary tuberculosis, caused by the tubercle bacillus, and it is transmitted mostly through the alimentary and respiratory tracts. The bacillus hibernates in a constantly dormant state in more than half of the human race, with no effect on most peoples' lives, although in its aroused virulent state it has become the most infectious disease and killed hundreds of millions of people throughout the world. The man rightfully credited with contributing most to the virtual stamping out of this dreadful sickness, which is making a comeback in a different strain, was the German bacteriologist Robert Koch (1843–1910), who isolated the TB bacillus in 1882. Moreover, the pasteurization (named after the great French chemist Louis Pasteur, 1822–1895) of infected milk has helped in eradicating millions of potential cases of tuberculosis. Koch also identified the comma bacillus, isolating the microorganism that caused cholera in 1884, and he found means to combat the disease of rinderpest.

☞ A 97.

The high glucose level in the mouth of a diabetic patient tends to thicken the walls of blood vessels and these thickened walls often prevent the gums from getting all the nourishment and oxygen they need. Bacteria find it easier to infest these areas. The thickened walls of the blood vessels also slow the gums' waste removal, which makes it more difficult to resist periodontal disease and infection. Careful brushing and flossing and four dental checkups per year can help avoid great periodontal harm.

☞ A 98.

Although the function of these drugs varies little, their origins go some way to explain their differences. Medications developed from molds and fungi are those antibiotic drugs that work by interfering with the production of whatever nourishment disease-causing bacteria require to do their damage, thereby inhibiting their growth. Antibacterials derived from chemical sources for the most part work by breaking down the cell walls of harmful bacteria.

☞ A 99.

If the spinach and chard—greens high in oxalic acid—are eaten with rice, the absorption rate of calcium is greatly increased. The oxalic acid is absorbed by the rice and thereby hastens the vegetables' release of calcium into the bloodstream. Other leafy green vegetables, such as collards, kale, and mustard greens, contain lower levels of oxalic acid and so are good calcium sources by themselves.

☞ A 100.

What would have happened? You wouldn't read these lines because you wouldn't exist. The first three neurochemicals produced in the brain are all natural amphetamines, and especially PEA produces feelings of elation and euphoria. When the brain, mostly through vision, becomes aware of a sexually desirable partner, it releases these neurochemicals, generating a euphoria of desire. However, as with any amphetamine, tolerance to PEA is built up in our blood and more and more PEA is required to sustain the amorous thrill. About three years after the initial encounter, the body can't generate sufficient PEA and the other neurochemicals to keep the sexual attraction at the same high pitch, and endorphins, which are chemically similar to morphine, enter the brain, replacing the feelings of continuous passion manufactured by PEA with calmer influences that result in sheer comfort with a loved one. Of course, the brain's pituitary gland still continues to secrete, in small amounts, the chemical called oxytocin, which mercifully will sustain, among other biological duties, the pleasurable feelings experienced during stimulating lovemaking and may possibly enhance or even instigate orgasm. Consequently, all five chemicals are imperative for the continuance of the family and the human race. By the same token, once the neurochemicals are psychologically put on hold, sexual attraction dries up and romantic love may come to a premature end.

☞ A 101.

During his Viennese student days, Semmelweis (1818–1865) became interested in puerperal infection (childbed fever), which was the great killer of young mothers throughout Europe, with a mortality rate of up to 25 percent. In the 1840s he discovered that students and doctors who came directly from the dissecting rooms to the maternity wards carried the disease from the corpses to the healthy mothers, infecting them. Semmelweis instituted washing in maternity wards with soap and water, and later ablutions with chlorinated lime, and the high maternal mortality rate dropped to below 1 percent. Even so, he was stymied at every step of his work by the medical establishment, and only his final act, sacrificing his own life by infecting a self-inflicted open wound with puerperalism, made him a martyr and savior of untold millions of mothers-to-be. However, when our question also mentions Oliver Wendell Holmes, we do not refer to the great American Supreme Court associate justice from 1902–1932, but to his father (1809–1894). The senior Holmes was a humorist and poet, but foremost a physician, lecturing at Harvard for thirty-six years. In 1843, just about the same time Semmelweis warned his doctors about the danger of spreading childbed fever to mothers-to-be, Holmes wrote his most distinguished medical work, *The Contagiousness of Puerperal Fever.* He, too, warned that the disease could be spread from patient to patient by the obstetrician, and he, too, met with vigorous opposition from leading obstetricians in the American medical establishment. Like Semmelweis, Holmes did not give up and was finally vindicated, although he did not have to sacrifice his life to prove his belief.

☞ A 102.

Virtually none. Leukemia has two main varieties: myelogenous and lymphatic. This cancer does not destroy red blood cells but is characterized by an abnormal increase of leucocytes (the *white* or colorless nucleated cells of the blood) in the tissues of the body and by a corresponding increase in the displacement and loss of red blood corpuscles. The increase in white cells in leukemia patients can produce up to one hundred times the normal count. However, in cases of acute leukemia the white cell count may actually be lower than normal, meaning that the number of platelets (minute blood particles involved in clotting) is decreased (thrombocytopenia). Symptoms of the disease are manifested by an abnormality in the bone marrow, although the lymph glands, liver, and spleen are also frequently affected.

☞ A 103.

Something else. Ptyalin is an alpha-amylase, an enzyme that accelerates the hydrolosis (the chemical process of decomposition) of starch, etc., and is found in the saliva of people and many animals. If you let a cracker dissolve on your tongue, for instance, ptyalin lets the starch of the cracker disintegrate into the individual substances that compose it, mostly sugar molecules. These sugar molecules or sugar links break up in your digestive juices before they enter the intestines and finally escape through their walls into your bloodstream.

☞ A 104.

In the first place the HIV virus is never dormant but produces millions of copies that can wait up to ten years in the victim's lymph nodes until much of the body's immunological system is put out of action. That's when it invades the blood. Hiding in the body's vital lymph glands or nodes—localized masses of tissue distributed along the lymphatic vessels—makes it virtually impossible to eliminate the microscopic viruses. AZT has been clinically proven to help those with AIDS but only for the first several months before and after life-threatening symptoms appear.

☞ A 105.

The 99 percent is made up of calcium phosphate ($Ca_3[PO_4]_2$) and calcium carbonate ($CaCO_3$), and this high percentage of these two compounds of the chemical element calcium is found mainly in our bones and teeth. (The blood and organs contain the remaining 1 percent.) Calcium keeps teeth and bones strong, but it also helps the body utilize iron and expedites the passage of nutrients in and out of our cells.

☞ A 106.

About two hours after inserting a gene—the gene that regulates the construction of luciferase and thus makes fireflies glow—into diseased tissue samples taken from TB patients, the infected cultures will be set aglow. A number of TB-fighting agents are then combined with the cultures to determine whether a particular strain of tuberculosis is susceptible to the drugs. If the light goes out or is dimmed, the TB strain is susceptible, but if the culture continues to shine, it means that strain is resistant.

☞ A 107.

The chief culprit of loose teeth in adults is not so much the root of the teeth but the surrounding gum, which recedes because of disease, old age, faulty diet, and unnecessarily brisk brushing. The result is that the dentin—the ivory tissue that forms the body of the tooth under the enamel—becomes increasingly exposed, and that freshly exposed part is more porous than the older tooth surface. The now unprotected dentin produces new conduits to the nerves inside the tooth's pulp and to the affected teeth's roots. Frequently, good dental care can prevent further damage to the teeth if gum problems are caught in time and precautionary measures are carefully followed under a dentist's supervision.

☞ A 108.

Microbiologists are now convinced that some bacteria are resistant to these miracle drugs because of the mercury used in dental fillings. A complicated process causes the inhaled mercury vapor to enter our bloodstream, lungs, and intestines, making bacteria to a large extent impervious to many of these antibiotics after our cells change the mercury into mercury ion and transports it to our intestines, from where it is excreted. Not that the bacteria feeds on the mercury—about half of the dental amalgam—but for some unknown reason mercury reinforces the bacteria against some antibiotics.

☞ A 109.

The removal of worn-out red blood cells in any healthy human being is the job of the macrophages ("big eaters") present in the liver and spleen. But the bone marrow alone gives birth to about 140,000 new red blood cells every minute, and their main function for about four months is to carry oxygen to each part of the human body. At the end of this time the macrophages eliminate the old cells, making room for fresh ones.

☞ A 110.

None of these. It is cataracts that develop in about 400,000 senior citizens (and some young patients) in an average year. In 1991, for example, 1,350,000 cataract operations were performed, costing the Medicare program alone around $3.4 billion.

☞ A 111.

Based on the findings of their soil samples, they discovered the antibiotic Aureomycin, the first of the tetracyclines, which are used to treat Gram-positive and Gram-negative bacteria as well as a wide range of infections, such as certain viruses, rickettsiae, and some protozoal parasites. Duggar successfully isolated and fully described Aureomycin as late as 1948. It is the trademark for the preparation of crystalline chloretracycline hydrochloride. A tetracycline is an antibiotic that is derived from a bacterium and is used to treat infections. Gram-positive means to retain a stain by alcohol in Gram's method, a primary characteristic of certain microorganisms. Gram-negative means to lose stains, or to decolorize them with alcohol in Gram's method of staining.

☞ A 112.

A thrombosis is the formation of a blood clot in the circulatory system, and an embolus is the occasion when a thrombosis disengages itself (including an air bubble) and blocks one or several blood vessels. These two terms were coined in 1848 by the German pathologist, anthropologist, and political leader Rudolf Virchow

(1821–1902), probably the world's best-known medical doctor during the last third of the nineteenth century. (In case you're planning to challenge that last statement, Louis Pasteur [1822–1895] was a chemist and microbiologist.) Among Virchow's many accomplishments were the discovery of myoma, myxoma, and glioma, and the redefinition of sarcoma and melanoma. A pioneer in the field of public health, Virchow fought for better hospitals, schools, and improvement in sewer systems. He aided archaeologist Heinrich Schliemann (1822–1890) at Troy and was keenly opposed to the policies of German chancellor Otto von Bismarck (1815–1898).

☞ A 113.

The physician was the pioneer of vaccination, Edward Jenner (1749–1823). He realized that milkmaids never caught smallpox because once they contracted the relatively harmless disease of cowpox from the cows they milked they gained immunity to the related, but much more pernicious disease smallpox. Moreover, Jenner coined the word *vaccination* from the Latin word for cowpox: *vaccinia.* To test his theory, he infected an eight-year-old boy in 1796 by inoculating him with matter from a cowpox-diseased cyst from a milkmaid. This built up the necessary resistance in the boy's bloodstream, because when Jenner inoculated the child with smallpox two months later, the boy remained uninfected.

☞ A 114.

England's Thomas Robert Malthus (1766–1834) warned the world, and especially England, of overpopulation in his second revised 1803 book, *Essay on the Principle of Population.* There he claimed that populations increase in geometric ratio and food only in arithmetical ratio. He predicted that war, disease, and famine were going to be necessary checks on population growth. Has the world listened to Malthus since then? Let's see: Between 1800 and 1920, the world's population grew from around 1 billion (a thousand million) to 2 billion people, and to 3 billion in 1960. At the end of the twentieth century there will be at least 6.25 billion and by 2025 around 8 to 9 billion people, with about 95 percent of the increase in the most impoverished parts, the third world, according to Yale professor Paul Kennedy. Between 2025 and 2035 India will have more inhabitants even than China, around 2 billion, and later next century Mexico will have as many as 180 million inhabitants, twice as many as today. Plainly, the world has not taken Malthus's theory to heart. Malnutrition, disease, crime, and war, particularly in the Third World, will be a natural response to hold further abnormal population growth in check while many religious faiths will support the bulging population increase. The United Nations estimates that by the middle of the twenty-second century, the world's population will have risen to 11.5 billion people (twice as many people as today), a comparatively small increase from the population estimate of what is projected for 2025 (9 billion people). The U.N.'s figure, consequently, is highly optimistic.

☞ A 115.

The irony was that the German doctors at the Pomeranian military hospital had developed a technique to successfully treat their blinded patients—including Corporal Hitler—and this medical treatment was based on theories developed a short time earlier by Eugen Fraenkel (1853–1925), the discoverer of the bacillus that seeks nourishment from exposed human body parts subjected to chlorine gas (*Gasbrand Bazillus*). Fraenkel was an eminent German pathologist and bacteriologist who was Jewish. After Hitler came to power in 1933, Fraenkel's monument and gravemarker were removed or destroyed. Today a street in Hamburg is named for Fraenkel in honor of his many medical achievements, and his bust has been re-installed in Hamburg's Eppendorf hospital, where he practiced medicine until his death.

THE HOME PLANET

☞ A 1.

The answer is no. But not because the Siberian temperatures are incapable of going below –70°—the record dipped to –90° F on February 6, 1933, in the Siberian village of Oymyakon (population c. 4,000)—but because mercury is not used in those icy regions. Mercury freezes at a point between –39° and –40° C and F, so the Siberians use alcohol in their thermometers, and alcohol does not freeze until a temperature of –114° C is reached. It is reported that the temperature in Oymyakon recently dipped to –98° F—unofficially.

☞ A 2.

No, it's not Mount Everest. Mount Everest is indeed the highest mountain if you measure it from a base of *sea* level to its summit, 29,028 feet (8,848 meters), and rising. But the question specifically asked that the highest mountain be measured from its *base* to its summit. In this tricky case the base of the extinct Hawaiian volcano Mauna Kea is actually 19,087 feet *below* sea level, and from there it rises another 13,784 feet (c. 4,200 meters) above sea level, with a total elevation of 32,871 feet. It beats Mount Everest by about 3,843 feet. As a matter of fact, because of its elevation high above clouds, atmospheric moisture, and artificial lighting, Mauna Kea is an ideal site for infrared astronomy. The U.S. installed its first and the world's biggest infrared telescope on this Hawaiian site in the 1970s.

☞ A 3.

The island is Bora-Bora, part of the fourteen Society Islands of French Polynesia in the South Pacific. It exports fruit, tobacco, and mother-of-pearl. The island is the exposed tip of an extinct volcano. Over hundreds of thousands of years it was built up by uncountable basaltic lava flows. The coral reefs virtually encircling the island are made up of deposits of calcium carbonate that colonies of minuscule coral animals erected over eons. It is only the narrow gap in the encircling reef, Teavanai Pass, that enables vessels to enter the sheltered lagoon. In fact, this southernly lagoon, Turaapuo Bay, is the main crater of the volcano that formed the island.

☞ A 4.

Rising sea temperatures will have profound and tragic consequences for life in our oceans. Quite apart from natural climatic disturbances in many parts of the world, fish, plant life, and coral growth will die off due to increased cloud cover. In addition increased rainfall will dilute the salinity of lakes and shallow ocean waters. The combination of all these and subsidiary pollution factors is bound to decrease life in the oceans.

☞ A 5.

You are actually living in the Cenozoic Era right now. It started about 65–70 million years ago at the beginning of its subdivision, the Tertiary period. It is marked by the rapid evolution of birds, mammals, whales, apes, and grazers as well as of shrubs, trees, and grasses, although there was relatively little change in the invertebrates. The Tertiary period, which preceded our own Quaternary period, was marked by the formation of mountains, such as the Alps, the Himalayas, and the Caucasus and the eventual dominance of mammals and, later, the human race on land.

☞ A 6.

There have never been hurricanes by those names because hurricanes are named annually in alphabetical order, and there have seldom been more than fifteen of these serious storms in the Atlantic in any given year this century.

☞ A 7.

First of all, both a typhoon and a hurricane are types of storms referred to as *tropical cyclones.* Second, hurricanes now are considered storms with a mean velocity of over 75 mph (c. 120 kph) that originate in the Atlantic Ocean. Originally these storms were said to develop in the West Indies, around the Lesser Antilles, and in the Gulf of Mexico. Typhoons are tropical storms starting out in the western North Pacific area, although they can move as far as China, Japan, and even hit the coasts of Siberia and Korea. A hurricane in Kansas? Not likely. What Dorothy experienced in her dream was a tornado. Hurricanes exhaust themselves long before they could reach as far into the interior of the United States as Kansas. And when it comes to a monsoon, it's a series of winds striking southwest Asia and the Indian Ocean. Monsoons have also been known to hit the Gulf coast of the United States (rarely), and northern Australia. Southeast Asia from the Arabian Sea to China have been hard hit, too. Heavy rainfall frequently accompanies these storms.

☞ A 8.

It is the second largest lake in South America, Lake Titicaca, which is divided between Bolivia and Peru on the Altiplano, the upland plateau between the eastern and western Cordilleras of the Andes. Its elevation is about 12,500 feet (c. 3,800 meters) and the people living there (Quechuan, Aymaran, and Uran Indians) differ from those of lower regions in that they have more red blood corpuscles as well as larger spleens, hearts, and lungs to enable them to survive in the thin, oxygen-poor air. The lake's area is about 3,150 square miles (c. 8,250 square kilometers), but it was larger in the past, when the Aymaras were the first people to plant potatoes as a crop. Covering an area between 5,000 and 6,000 square miles, Lake Maracaibo in northwestern Venezuela is the largest lake in South America.

☞ A 9.

It becomes marble.

☞ A 10.

Up until the summer of 1992 it was believed that an even more disastrous catastrophe was visited upon our planet 370 million years ago, during the late Devonian period. It is believed to have wiped out three-quarters of all marine species on Earth. Tiny glass beads merely 0.3 mm (0.004 in.) long, called microtektites, were found in underground sediments in Belgium. Today scientists are of the opinion that these tektites originated from silicon and other minerals melting and finally cooling as a result of a stellar impact or a gigantic volcanic eruption. But in this case the chemical composition leads geophysicists and astronomers to believe that it was an impact, and this is supported by two giant craters that are about 370 million years old and were found in Sweden and in Quebec. However, a year later, in late summer 1993, Dr. Virgil L. Sharpton of the Lunar and Planetary Institute in Houston, wrote in *Science* that the speeding object that struck Earth from outer space about 65 million years ago in the present Gulf of Mexico and the northern tip of the Yucatán Peninsula caused a crater that was about 185 miles in diameter, which today is buried about a mile beneath the Earth's surface. The speeding object striking Earth from outer space could have been as huge as ten miles across. The energy released by this colossal, unimaginable impact is estimated to be the equal of about 300 million hydrogen bombs, with each of these bombs being about seventy times more powerful than the atomic bomb dropped on Hiroshima. Dr. Sharpton believes that the energy released by this impact vaporized thousands of cubic miles of this planet, blocking out the sunlight for such a long time that it hastened the demise of the dinosaurs. Today this catastrophe is also believed to be the most destructive force to have struck this planet since any form of life began between three and four billion years ago.

☞ A 11.

All three were among the dinosaurs that lived in the Mesozoic Era. The first two had no teeth, but the oviraptor had two tiny ones. These reptiles probably survived by eating insects, eggs, some plants and, possibly, small animals. The more famous *Tyrannosaurus,* on the other hand, had most likely the biggest teeth of any dinosaur—six to eight inches long. It was also the largest meat-eating land animal to roam the planet Earth. But for some reason the best known of the dinosaurs is the *Brontosaurus,* which really is not an accepted name any more. This giant, a sauropod that evolved from a flesh-eating to a plant-eating animal, is more appropriately referred to today as an *Apatosaurus,* which could be as tall as 70 feet (21 meters), surpassed only by the *Seismosaurus,* which reached twice that height. They flourished in the Jurassic period, which ended about 137 million years ago.

☞ A 12.

It is almost impossible to tell since most fossils have been lost forever. But paleontologists estimate that we have records of only one species in every 10,000 that has ever lived and then perished throughout the ages. That is *species,* not animals. Virtually all the surviving species, at least their skeletons or fossilized parts of them, have survived only because they were buried in limestone and the clayey ooze at the bottom of the continental shelves.

☞ A 13.

Sometimes drumlins are also referred to as drums. They are not musical instruments, however, but long narrow ridges of drift or alluvial formation, resembling oval-shaped hills. Hundreds of millions of years ago they were formed by flowing glacial ice. These glacial elevations are overgrown with grass or woodlands today.

☞ A 14.

Not the carbonic acid usually found in most caves, but sulfuric acid, which would have proved fatal to anyone present at the time of the cave's formation. The famous speleologist Carol Hill, who is associated with the University of New Mexico, discovered that sulfuric acid carved out these caverns as a result of a reaction between oxygen that was dissolved in groundwater and hydrogen sulfide that emanated from well below the caves' surface. In most other caves carbon dioxide has mingled with rainwater to produce carbonic acid, which disintegrates the limestone bedrock of the caves. The resultant excavations, which created a labyrinth of subterranean passages, formed new underground streams.

☞ A 15.

The first transatlantic cable was laid in 1858 between Newfoundland and Ireland.

☞ A 16.

Benjamin Franklin was the scientist who first charted the Gulf Stream. It can be as much as eighteen degrees warmer than the icy Atlantic waters through which it passes.

☞ A 17.

Don't dispute the fact! It's true. Karst really takes up that much space underneath Mother Earth's surface. What happens is that underground caves are often etched out of limestone by acidic streams. This weak acid dissolves bedrock, forming underground rivers. These limestone terrains—static underground pools or flowing waters—are referred to as karst. And although Slovenia does not make up even 1 percent of the earth's surface, karst is actually named after Slovenia's Karst limestone plateau which stretches east of the Adriatic, making up one of many regions where karst is prevalent.

☞ A 18.

In the Book of Genesis, Abraham impregnated his wife Sarah's handmaiden, Hagar, when Sarah at first proved unable to conceive. The birth of Ishmael raised enmity not only between him and Abraham's legitimate son,

Isaac, but the conflict between those two has raged every generation since. Muslims believe that it was Ishmael, not Isaac, that God asked Abraham to sacrifice and that Ishmael helped Abraham build the Kaaba in Mecca.

☞ A 19.

Hardly. The Three Mile Island nuclear incident took place in Pennsylvania in 1979. The Three Mile Limit law was first written into law toward the end of the eighteenth century. It refers to the limit of a country's geographical jurisprudence over the waters adjacent to its shore. In 1793, three miles was adopted by the United States, by Great Britain in 1878.

☞ A 20.

In a 1986 article in the science magazine *Nature,* a trio of biologists suggests that a woman they named Eve lived between 140,000 and 280,000 years ago in sub-Saharan Africa and that she was the ancestor of *Homo sapiens* on Earth today. She wasn't the ancestral mother of all humans; other female ancestors of the genus *Homo* lived in Africa nearly four million years ago, reproducing even before her, and these ancestors have modern descendants. But their Eve appears in everyone's genealogy. This conclusion was drawn by studying mitochondrial DNA (mtDNA), which is the DNA outside the cell nucleus and inherited only through the mother. It is easily traced since it never mixes with paternal genes as it is passed from generation to generation. A 1991 study revealed that the greater genetic diversity of living Africans proved that they are the product of the longest evolutionary lineage. At the same time, Dr. Alan Templeton, a geneticist of Washington University in St. Louis, maintained in a 1992 *Science* article that scientists may never resolve the issue of modern human origins and that one cannot deduce the evolutionary picture of a species from one strand (mtDNA) of genetic evolution. Some other molecular biologists agree with the latter theory, maintaining that the idea of a single place of origin (which definitely does include Africa) can be disputed and that the evolution of modern humans over 200,000 years ago occurred almost simultaneously in many places due to a climate and flora favorable to the genus *Homo*.

☞ A 21.

In 1913, the United States Coast Guard organized and conducted the International Ice Patrol with all the most up-to-date scientific paraphernalia. Since that date, just one ship has been reported to have struck an iceberg, with a single fatality. It happened during World War II, when the patrol had been temporarily discontinued. Recent oil-tanker mishaps have not resulted in fatalities by drowning when collisions with icebergs occurred (an extremely rare event).

☞ A 22.

Even though many astrophysicists maintain that asteroids extinguished the dinosaurs in the Cretaceous period (65–70 million years ago), a few paleobiologists insist that *all* mass extinctions resulted from the impacts of asteroids or comets, such as the space rock with a diameter of about twelve miles that smashed into

Earth 250 million years ago. Other scientists claim that nine out of the ten greatest mass extinctions could be credited mostly to continental floods of volcanic basalt. A few scientists believe that certain rocks known as tillites were caused by an extra-terrestrial collision, while other researchers ridicule this idea since tillites are so common on this planet. To add to the confusion, other researchers say that the precious metal iridium, which belongs to the platinum group of elements, is more abundant in meteorites than in terrestrial rocks. The high content of iridium in strata laid down at the Cretaceous–Tertiary periods provides strong evidence that extraterrestrial impacts did take place at those epochs. Another group of scientists declare that asteroids helped to break up the supercontinent Gondwanaland 250 million years ago causing volcanic eruptions that led to the Permian period of the greatest mass extinction on Earth.

☞ A 23.

Evolution had already been discussed by Jean Baptiste de Lamarck and Darwin's grandfather, Erasmus, before Charles Darwin was born. Charles Darwin started to write his *Notebooks on the Transmutation of Species* in July 1837 and in 1844 finished a treatise on the theory of evolution by natural selection, which he showed to the botanist, Sir Joseph Dalton Hooker. He also summarized his theory in an 1857 letter to the American naturalist Asa Gray. When Darwin received Wallace's article detailing the simultaneously developed, identical theory in 1858, he immediately placed the article, and a few days later, his own 1844 treatise and a copy of the letter to Asa Gray, in the hands of the geologist Sir Charles Lyell (1797–1875) and the botanist Sir Joseph Dalton

Hooker (1817–1911), who were both familiar with Darwin's earlier work. These two men decided to have the Wallace article and the Darwin treatise read as a joint paper at the Linnaean Society in London on July 1, 1858. Of overriding importance however, are Wallace's own words in the preface to his work *Natural Selection* (Darwin's term), in which Wallace acknowledges that Darwin had been at work on *On the Origin of Species* long before he formulated his own ideas on the subject and that Darwin alone of all men was best fitted for the great work he had undertaken and accomplished. Only posthumously did the Royal Society recognize Darwin's preeminent contribution to science by estabishing the Darwin Medal. Its first recipient was the co-originator of the theory of evolution by natural selection, Darwin's friend Alfred Russel Wallace.

☞ A 24.

Nobody knows for sure what caused life. But many astronomers and other scientists believe that the arrival of comets on our planet helped to trigger the formation of life here. In 1985–86, Japanese, Russian and European space probes chemically broke down and classified the cometary grains of Halley's Comet, which orbits the Sun every 76–77 years in the opposite direction to the planets; it is the most conspicuous and brightest of the periodical comets and will make its next appearance for Earthlings in 2061. It was discovered that the basic chemistry of the comets crashing on Earth contained particles that consisted of carbon, hydrogen, nitrogen, and oxygen. These are exactly the chemicals that make up almost every living thing. As cometary dust and meteorites kept hitting our planet over billions of years, they are believed to have transferred these life-bearing chemicals to Earth from outer space. A more recent (1993) school of molecular biologists, astrophysicists, and astronomers, however, believes that interplanetary dust particles—outweighing the fall of meteorites on Earth by 100,000 to 1—carried tiny particles of organic compounds that originally helped to produce life on our planet about four billion years ago.

☞ A 25.

His name was Ptolemy (Claudius Ptolomaus). He lived in the second century A.D., and not much is known about his life. Most of his work was carried out in Alexandria. He was one of the great astronomers and mathematicians of his age, wielding tremendous influence on future generations for almost 1,500 years, although much of his work was based on that of Greece's Hipparchus (c. 190–120 B.C.). Ptolemy's *The Mathematical Collection* later became known as *The Great Astronomer,* or *Almagest.* Until the age of Copernicus, his thesis that the Earth is the center of the universe prevailed. He suggested that our planet was a globe, not flat. He prepared a calendar, giving the risings and settings of the stars in morning and evening twilight with weather indications. He also wrote several first-rate books on mathematics, geometry, and optics.

☞ A 26.

In California. It is the part of the 800-mile-long San Andreas Fault, on which Los Angeles is built. Los Angeles could thus, once more, become the pride of beachfront surfers—next time, however, in Alaska. In this particular instance, two sections of the Earth's crust, or plates, slide unstoppably across the sticky, incredibly hot rock mantling the molten outer core of our planet until the western or Pacific plate of the San Andreas in its push north could reach the outer confines of Alaska.

☞ A 27.

When heated, the bituminous matter called kerogen decomposes to yield shale oil. The sedimentary rock is referred to as oil shale, which is not just a lithologic term (lithology explains the character of rock formation), but also an economic term. In this respect, the definition of lithology relates to scientific and economic consequences of yielding oil from rock formations. Because of its high yield of oil, this source of liquid fuel is cheaper and more plentiful than any other except natural petroleum.

☞ A 28.

By neither. The English philosopher Herbert Spencer (1820–1903) most likely used the word *evolution* for the first time in "biologically correct" essays dating back to 1852, describing the idea of a general process of production of lower and higher forms of life. It is true that Darwin's teacher and mentor, Scottish geologist Sir Charles Lyell (1797–1875), used the term in the 1830s but in a less general sense. However, it was Spencer's definition of the word that caught on. Oddly enough, the word was rarely used by Darwin, who also used the term *survival of the fittest,* but only once in *On the Origin of Species* (1859). That phrase was also coined by

Spencer and picked up by Darwin as well as by Wallace (1858). The theory of the *survival of the fittest* was expounded in Spencer's essays "The Development Hypothesis" (1852–1854) and "Progress: Its Law and Cause" (1857), and used in his 1864 book *Principles of Biology.* Darwin and Russel Wallace believed that evolution could only be attributed to natural selection, whereas Spencer at first was convinced that the inheritance of acquired characteristics were the ultimate principles of the evolutionary process. Later Spencer accepted Darwin's and Wallace's view that natural selection was one of the causes of biological evolution, but he clung to his primary observation that any natural force continuing to wield its superiority must also evolve and change. This was Spencer's main observation regarding biological development. It should be emphasized, however, that Darwin discussed the theory of *evolution* in an unpublished essay written between 1837 and 1844, and Jean Baptiste de Lamarck (1744–1829) claimed that evolutionary characteristics are inherited, but Darwin's work had not yet been published in 1852, and Lamarck's interpretation of the evolutionary process (which he preferred to call "transformation") was later partially discredited.

☞ A 29.

Not as you may think, by stretching its neck to reach foliage in tall trees. It is because of the giraffes' mating with antetypes that had a longer neck and this species outlived those with shorter necks. This is an example of the natural-selection theory as propounded in the work of Alfred Russel Wallace and Charles Darwin in the mid-nineteenth century.

☞ A 30.

He is commonly referred to as the Java Man because that is the location (near Trinil) where his bone fragments were found by Dr. Eugene Dubois in 1891. He belonged to the now-extinct species *Homo erectus,* which lived in Europe, Asia, and Africa. While the Java Man, disparagingly referred to as "ape-man," existed for about half a million years shortly before the major ice ages were about to begin, the species classified as *Homo erectus* is known to have lived for about 1.8 million years. It was actually Java Man (*Pithecanthropus erectus*) who first used fire and developed many stone tools. This species became extinct about 400,000 years ago. *Homo sapiens* were known to have been around 200–400,000 years ago, but anthropologists are not certain whether we evolved from *Homo erectus* or replaced the latter altogether with Neanderthals and the later Cro-Magnons, whose cycle came to a close about 40,000 years ago, and who were followed by or interbred with us: *Homo sapiens sapiens.*

☞ A 31.

Because iridium is a common metal found in some meteorites and asteroids and traces of it were detected in layers of clay and rock that date back about 65 million years—approximately the time the dinosaurs became extinct.

☞ A 32.

When the Cretaceous period ended, a huge object from outer space slammed into the Yucatán. Today it is almost certain (although still doubted by a few paleontologists) that this cataclysmic impact was the main contributing factor to the mass extinction of dinosaurs. The 185-mile wide crater, long buried by sediment, is the most likely ground zero. Melted rock from the crater itself is 64.98 million years old (other asteroid strikes occurred in Europe, New Zealand, and the Atlantic). Not only the melted rock but also glassy debris, known as tektites, were created by the fiery impact and have been located on the relatively nearby island of Haiti and in northeastern Mexico. The resulting pall of dust blotted out the sun, which affected the growth of vegetation—part of the food chain of dinosaurs. But the main cause for their extinction probably was the vaporized limestone of the Chicxulub crater (the exact location of the Yucatán peninsula where the impact took place), since this vaporization would have proven highly noxious with its release of incredibly huge amounts of carbon dioxide, killing off the dinosaurs, if not in addition all plant life on Earth. In 1992, Dr. Walter Alvarez, a geologist at the University of California, and before him his father, the late physicist Luis Alvarez, as well as Dr. Carl C. Swisher, 3d, at the Institute of Human Origin in Berkeley, and their colleagues, identified the meteorite's impact age with an advanced method called argon-argon dating. Oddly enough, there is evidence now that the dinosaurs would have perished anyway, about 65 to 73 million years ago, with about 70 percent having become extinct already during that crucial period.

☞ A 33.

About three-quarters of the Earth's landmass is made up of sedimentary rock. Detritus builds up at the bottom of lakes, ponds, rivers, and oceans. Over millions of years pressure from new layers of material forms a sort of a natural goo, converting the material into sedimentary rocks. Layers of clay and mud turn into shale and are typical representatives of sedimentary rocks. More than 20 percent of the Earth's continental surfaces are made up of igneous rocks. They are the result of frozen molten magma that pour out of erupting volcanoes. Igneous rocks' best known manifestations are granite and pumice. Only a small percentage of land on the Earth's continental surface is made up of metamorphic rocks. These are changed over millions of years from their original state to a new substance by the constant pressure of geological activities, such as astronomical temperatures and chemicophysical side-effects. Limestone ending up as marble is a typical example of metamorphic rock transformation. However, limestone itself has been formed partly over eons from the indestructible parts of marine organisms, so it can be considered a kind of sedimentary rock.

☞ A 34.

With an elevation of about 12,400 feet (c. 3,800 meters) above sea level, La Paz is the highest capital city on Earth. Tibet is the largest area of high land in the world, with an average altitude of about 16,000 feet. Its capital, Lhasa, is 11,830 feet above sea level.

☞ A 35.

Geologists refer to the "Cambrian explosion" when they discuss the period of 590 million years ago when physical structures of living creatures generally began to evolve into skeletons, the Cambrian Period. Invertebrate animal life appeared in this first period of the Paleozoic Era, and marine life was widespread, too. Dating from this period are the earliest fossils of creatures with hard shells, such as trilobites, a group of articulates with a three-lobed body. Their hard parts and skeletons became fossils, and because there is such an abundance of them dating back to approximately 85 million years into the Cambrian Period, scientists often refer to this time as the "Cambrian explosion." Of course, living things existed before the Cambrian Period, but few fossils remain from that time since harder and more durable bone structures had not yet fully developed. Life existed only in the oceans at first, and for at least one hundred and fifty million years; it did not venture onto land until about 430 million years ago, during the Silurian period. In fact, plant life preceded animal life to coastal land regions. The first fully developed land animals were believed to be scorpions.

☞ A 36.

The correct answer is Fahrenheit *and* Celsius. There is only one temperature at which the readings are the same on both scales: –40° C is the same as –40° F.

☞ A 37.

Oil. To start with, the medicine was known as "rock oil" or "Seneca oil," and it was rubbed on the body to treat open sores and burns. People had known for centuries about oil but had no idea how to dig for it. In 1859, oil extracted near Titusville, by Pennsylvania businessmen—prospectors—by drilling sixty-nine feet, was refined and turned into kerosene, which was used primarily to light lamps. Only toward the end of the nineteenth century did oil become the most sought-after fuel in the world, especially after the invention of the motorcar. Today, of course, the environmental movement challenges oil's dominance and its damage to living things. Yet in the early 1940s, higher-octane fuel helped British Spitfires and Hurricanes to outmaneuver the Nazi enemy with his Messerschmitts during the Battle of Britain, and later too. Oil and petroleum played an enormous part in both world wars and proved to be as coveted a treasure to Hitler and Hirohito as new land and grain. More recently, oil rights were the principal economic reason for fighting in Kuwait (1991).

☞ A 38.

By watching the water run out of the bathtub. If it drains counterclockwise, he is in the Southern Hemisphere, and if it drains clockwise he is in the Northern Hemisphere.

☞ A 39.

Both sites boast some of the biggest craters on our planet dug out by comets, meteorites, or asteroids. Located in Ontario, the Sudbury Crater—the bigger and older one—is about 1.8 billion years old and 300 kilometers (188 miles) wide. Less impressive, but still awesome, is the smaller Manson Crater near Iowa City, which is "only" 35 kilometers (22 miles) across and about 65 million years old. Geologists believe this crater was caused by a substantial bulk of the giant comet that may have been partly responsible for the extermination of the dinosaurs.

☞ A 40.

There is just one place in the world where they can be found living together on the equator: the Galapagos Islands, because the waters there are cooled by the Humboldt Current, which sweeps northward out of Antarctica.

☞ A 41.

The Akkadians' demise has indeed always been a mystery . . . until now. Soil samples in parts of Iraq and Syria, where the civilization flourished, were examined by Yale University archaeologist Harvey Weiss and his French–American team. They discovered that there had been a disastrous dry spell for about three centuries in the once-rich farming region. This long drought was the result of an incredibly powerful volcanic eruption that blanketed the entire region with an almost impenetrable layer of ash. In the summer of 1993 a *Science* magazine article concluded that the drought caused a climate change, which was in turn responsible for destroying large parts of the agricultural life from India to Egypt.

☞ A 42.

It does indeed. The Earth's motion is perpetual. All artificial motion, such as movement resulting from man-made objects, like a pendulum, and chemical as well as physical terrestrial phenomena, is temporary. The

Earth's motion is irresistible. Resistance puts all other motions to rest.

☞ A 43.

Clarke proposed the idea of communications satellites that are stationary in certain locations above our planet. He referred to them as synchronous satellites, or just plain satellites. Between 1965 and 1967 they were placed in a geosynchronous orbit (moving at the same speed as Earth, so as to remain fixed above the equator) and became the indispensable and primary means of intercontinental communication.

☞ A 44.

Hardly. Magma is a deep-seated reservoir of molten rock that rises through a conduit to the outer crust of a volcano. Once it spills from the volcanic vents over the top of the volcano's crater it is called lava. Igneous rock is formed when the magma and lava cool in atmospheric conditions. Volcanoes can erupt often without warning, and the molten lava will destroy anything in its path. Driven by the force of the magma, lava can have a flow as deep as fifty feet and more. Nevertheless, some volcanoes explode violently with huge boulders but without a lava flow. In Hawaii some of the fluid basaltic magma and lava even reach temperatures of up to 1,300° C. A particularly violent volcanic eruption occurred in what is now Italy in A.D. 79, when Vesuvius erupted and thousands were killed under tons of lava, including the Roman scientist Pliny the Elder, who was among those suffocated by the poisonous fumes that reached as far as Pompeii and Herculaneum. The largest volcanic explosion in the modern era took place between Java and Sumatra when Krakatau (Krakatao) erupted on August 27, 1883, killing more than 36,000 people. However, the worst volcanic eruption with the most human fatalities was on April 5–7, 1815, when about 92,000 people were killed in Tambora, Sumbawa, Indonesia.

☞ A 45.

Not since the beginning of the 1990s. Ever since the invention of a process called polymerase chain reaction (PCR) in 1985, biochemists can take a tiny section of a DNA (deoxyribonucleic acid) molecule, the main constituent of a chromosome, which transmits hereditary characteristics, and make hundreds of billions of precise copies in less than a day. Biologists can then examine the genetic code of any mummified material, plant or fossil, especially when preserved intact in ice or amber, and analyze it in detail. One recent biological research study of an extinct termite has now determined that termites that were alive forty million years ago evolved independently from cockroaches, although they probably shared a common ancestor, never yet sighted by man.

NUMBERS AND FORMULAS

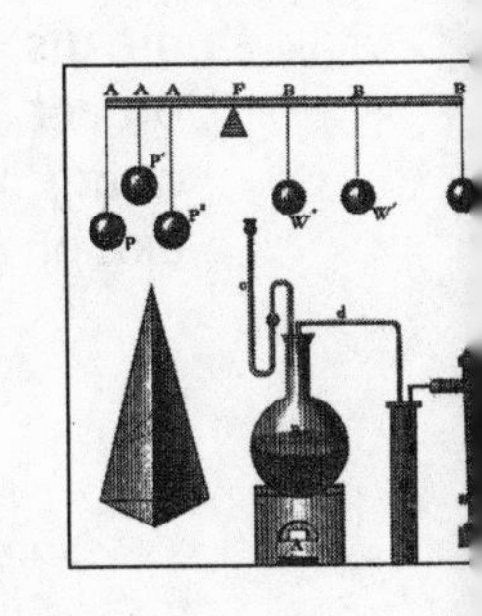

☞ A 1.

We have ozone galore on Earth already. But it is not the life-saving chemical found in the stratosphere. The ozone we have created on our planet is the life-threatening kind, whose principal element is found in urban smog. Breathing will always be impeded on Earth and will become more hazardous the higher the level of the ozone in our man-made smog. The combination of the ozone molecules with the effusion of the dirt we release into the troposphere (the hollow sphere of atmosphere surrounding our globe) results in the production of chemicals deleterious to our health, especially to that of our eyes and lungs.

☞ A 2.

The navigational term is dead reckoning. At one time a vessel's speed was determined by tossing a buoyant or floating object overboard and finding out how long it took for the length of the ship to pass the object in the water, provided the object remained static, or dead. Estimating, or reckoning, future positions of the ship by means of the speed required to pass this buoyant object (called a Dutchman's log) is known as dead reckoning. Using this estimated speed, the ship's location can be determined in mid-ocean.

☞ A 3.

Because in 1814 Fraunhofer, at twenty-seven, mapped 576 lines that he observed between the violet and red ends of the spectrum. These dark lines in the sodium spectrum of sunlight are known as Fraunhofer lines. He obtained the spectra of the stars by using a large prism through which the light passed (the prism was located outside the glass of his telescope), and was brought to a focal point in front of the eyepiece. Fraunhofer determined precisely, and for the first time, the optical constants of various kinds of eyeglasses that helped future opticians enhance the vision of the far- and near-sighted, and his technical achievements later aided scientists in the manufacture of achromatic lenses (which transmit light without decomposing it into its primary colors) and several optical instruments.

☞ A 4.

A wine connoisseur will tell you that the full bouquet of good wine can only be brought out after opening the bottle if you let the wine settle in your glass for several minutes. The wine must "breathe," because it is the abundance of oxygen in the air that dissolves some of the less robust molecules in the wine and produces the sought-after bouquet, the perfume exhaled by the wine. However, if wine is exposed too long to the air, the process of fermentation is halted, the wine turns sour, and chemical decomposition converts it to vinegar.

☞ A 5.

The alloy is bronze, which is a great deal harder than copper. The era named after this alloy, of course, was the Bronze Age. In the Middle East, especially in Turkey's Tarsus region, it dates back to 5000–1200 B.C. while in Europe it took place around 2000–500 B.C. Recent discoveries of bronze tools, however, prove that the Bronze Age actually began even earlier in the Far East, particularly in Thailand, than it did in the Middle East.

☞ A 6.

Moving, *not* motion, picture. It was not invented by England's Eadweard Muybridge in the 1870s, nor by France's Jules Etienne Marey in the 1880s, nor by Thomas Edison in 1894, as you may have thought, but by the English mathematician William George Horner in 1834. The contraption that displays it is a zoetrope, an optical toy in which people or animals on the inside of a revolving drum or cylinder are viewed through narrow slots that circle the drum. As the drum revolves the figure or figures on the inside circumference appear to be moving. Incidentally, the word *cinema* came into being in December 1895, when the French brothers Auguste and Louis Lumière called their first screening device Cinématographe, or *cinéma* for short.

☞ A 7.

A) Ken Olsen, who was the president of Digital Equipment Corporation. He made this observation at a World Future Society meeting in Boston in 1977, according to an interview he had with David H. Ahl in 1982.

B) The 1908 Nobel Laureate Ernest Rutherford (1871–1937) after he experimented with splitting the atom in 1933. He was the first to recognize the nuclear nature of the atom before World War I and is supposed to have named the nucleus in the atomic sense. He also discovered alpha, beta, and gamma rays.

C) Dr. Lee De Forest (1873–1961). He was the inventor of the Audion, which is used for a three-electron tube. Later he discovered that the triode (an electron tube with an anode, a controlling grid, and a cathode) could be utilized as an amplifier and oscillator. Improved by others, the triode made transcontinental wire telephony possible in 1915, and this in turn led to the development of radio broadcasting in 1920.

☞ A 8.

The number to be multiplied by 9, 18, 27, 36 through 81 is always 12,345,679. Multiply 12345679 by 9 and the total is 111,111,111. As the question states: each numeral is exactly the same as the one preceding and following it. If you add the numbers of these totals they will always amount to the figure by which 12,345,679 is multiplied; i.e., 9 × 12,345,679 amounts to 111,111,111 (nine ones). Multiply 12,345,679 by 18 and the total comes to 222,222,222 (nine twos amount to 18, the number by which you multiply). Multiply 12,345,679 by 27 and the result is 333,333,333 (nine threes amount to 27, the number by which you multiplied 12,345,679). Continue with this right through the multiplier of 81 and you will find that the result is nine nines.

☞ A 9.

The statement was made by Werner Heisenberg (1901–1976), the formulator of the Uncertainty Principle and the 1932 Nobel prize winner in physics for being the originator of quantum mechanics. Some scientists maintain that Heisenberg and his team had strived for atomic superiority in Nazi Germany while others claim they were scientific bunglers. Some scientific sources insist that he persuaded Albert Speer (1905–1981) and other high Nazi officials in June 1942 that it would take too long to learn how to produce the needed quantities of Uranium 235 to make the key atomic bomb material, firmly believing that Hitler would win the war soon anyway. Science historian Stanley Goldberg and Thomas Powers, a Pulitzer Prize-winning journalist, however, maintain in the September 1992 issue of *The Bulletin of the Atomic Scientists* that Heisenberg, who was at one time slated for assassination by the Allies, deliberately derailed the Nazi atomic program after a secret meeting with his mentor, Niels Bohr (1885–1962), in September 1941. In a couple of books they base their findings on the secret, recorded discussions among leading German scientists who had worked on the atomic bomb, during their August 1945 meetings in England's Farm Hall, a country estate near Cambridge, where the Allies had taken the scientists after the war. The irony is that, while in British captivity, they succeeded within a week in designing an atomic device similar to the one dropped on Hiroshima. Most important on those secret tapes, the German scientists, except Max von Laue, considered the war and the Nazi atrocities a nuisance, not a crime. Heisenberg was their scientific "Führer," who tried, vainly, to blackmail the British and Americans to guarantee the Germans control over their scientific institutes or he'd defect to the Soviets, which he didn't. He was a man of parts, politically blind, deceiving others and himself with his leadership.

☞ A 10.

Dry ice. As a matter of fact, carbon dioxide (CO_2), produced by the respiration of the animal kingdom, makes up between 0.03 and 0.05 percent of the air at sea level by weight. The irony is that once liquid carbon dioxide is compressed into dry ice, it does not melt back into carbon dioxide but vaporizes immediately, never returning to its original liquid state. This process is called sublimation.

☞ A 11.

The Scottish botanist was Robert Brown (1773–1858), and what he discovered was that particles move continuously in a fluid medium, such as liquid or gas. His theory is referred to as Brownian Motion or Brownian

Movement. It confirms the kinetic molecular theory that matter is composed of minuscule particles that never cease moving about; that matter is subject to impact from the molecules of the medium in which they move. All this was convincingly explained in 1905 by a scientist who later became an American citizen, Albert Einstein (1879–1955). Before moving to America, Einstein had been a German, then a Swiss citizen.

☞ A 12.

The upsilon meson, Y. A meson is an unstable nuclear particle with a mass between that of the electron and the proton. It can be negatively or positively charged, even neutral.

☞ A 13.

This time it was *not* a scientist who invented the name but the novelist James Joyce (1882–1941), in *Finnegans Wake* (1939).

☞ A 14.

John's English host would be within his rights to do so, and he would be both right and wrong. He'd be wrong because a billion (1,000,000,000) seconds in the United States comes to just under thirty-two years. And John's father could very well be thirty-two or more years old. But the Londoner is right, too, because in Great Britain (as well as in Germany and France) a billion reads 1,000,000,000,000, and he would rightly assume that John's claim that his American father was almost 32,000 years old is a foolish lie. Incidentally, what

Americans refer to as a billion, a one with nine zeroes, is called a milliard in the United Kingdom. (The British billion, a one with twelve zeroes, is called a trillion in the United States.)

☞ A 15.

The unit of power is horsepower (hp). The reason it is and was referred to as horsepower is to be found in the early evolution of the automobile, when the power of a vehicle was still likened to the power of a carriage drawn by horses. It must be emphasized, however, that different conversion factors apply in other countries. Incidentally, horsepower is also equal to 745.7 watts in the United States.

☞ A 16.

The first color photograph was taken well before World War I. In fact, it was taken the very year the American Civil War started, in 1861. This picture, of a bunch of grapes, is still exhibited at England's Cambridge University. The Scotsman who took it was one of the world's greatest physicists, James Clerk Maxwell (1831–1879). He predicted the validity of electromagnetic radiation, which in turn led to the discovery of radio and microwaves. Before Maxwell died of cancer in his forties he had contributed to every branch of physical science, including electricity, color sensation, magnetism, and optics, and his basic explorations have furnished future scientists with the fundamentals in telegraphy and telephony, while the four equations that he wrote shortly before his death unified the phenomena of magnetism and electricity, exactly as we know these physical agents today.

☞ A 17.

The spire is the top of the steeple. The steeple is that part of an edifice that is built on the roof or tower of a church and is meant to contain the bells. A steeple often tapers off into a spire, which is a separate part of the building's construction.

☞ A 18.

No such thing as bottomology exists. Topology is the branch of mathematics that concerns itself with the qualities and properties of space in contrast to the more refined analytic or geometric properties. Or put another way, it is a study in which the geometric properties and relationships are not affected by the distortion of a figure. The geometric forms remain invariant under certain transformations, such as stretching, bending, etc.

☞ A 19.

Strictly speaking, neither one nor the other. It's more complex than that. The wick, or string, by itself is not on fire. String on its own burns differently. (Try to light one.) But wax doesn't burn either. (Try to light the bottom end of a candle and it won't provide a flame.) So, what creates the flame? The heat of the candle's flame actually radiates down, warming the wax and melting it until it turns into a liquid pool at the candle's top. The wick soaks up the liquid wax and becomes increasingly hot as it approaches the flame, evaporating the liquid in the wick, and only the wax's vapor filters into the air. This vapor mixes with the oxygen in the air, and it is the wax's vapor mingling with oxygen that creates the candle's flame, *not* the wax or the wick by themselves. In fact, upon blowing out a candle, the wick's white vapor flows into the air, and if this vapor is lit with a match, the oxygenized vapor stream will return to its source, flowing down back to the wick and relighting the candle.

☞ A 20.

A total of fifty-nine minutes—because there are only fourteen rest periods, with none after the final, fifteenth round.

☞ A 21.

Stonehenge. Sarsen is the generic name given the blocks of sandstone that make up most of the ancient monument on Salisbury Plain in England. Sarsen is also called Druid stone or graywether. The 162 stones (weighing up to fifty tons) are probably remnants of eroded Tertiary beds, but they are believed to have been erected on Salisbury Plain, Wiltshire, between 1500 and 2000 B.C. Even though the well known surviving inner ring of Stonehenge with its upright pillar—the altar stone—today shows a bluestone as well as spotted dolerite and micaceous sandstone formation (eighty of these stones weigh up to four tons each), authorities are certain that they were erected after the sarsens because a number of them actually stand in fillings of ramps made of sarsens' sockets. It is even suggested that some of the dolerite horseshoe rocks were transported at a later stage from another unknown location. The original purpose of Stonehenge is still being disputed. It could have served as an observatory, a burial site, or a place of worship.

☞ A 22.

All these inventions and scientific discoveries were made by Robert Hooke (1635–1703), one of the great experimental physicists of all time.

☞ A 23.

Not according to *Science* magazine, which named nitric oxide its 1992 Molecule of the Year! NO is no no-no because it plays a major part in regulating blood-pressure, digestion, the nervous system, the immune system, and it is important for the proper maintenance of a number of our organs. In fact, without this former "poison" it would be almost impossible for the human race to procreate, since its presence is essential to the male erection.

☞ A 24.

It can be both. In the nineteenth century, the Scottish physicist James Clerk Maxwell (1831–1879) demonstrated the electromagnetic character of light waves. In this century it was discovered that photons, completely newly discovered properties, possess energy that seem to be packaged and that these photons resemble ordinary material particles. This was demonstrated by Einstein's analytical thesis on the photoelectric effect. Better known, of course, is the fact that light is the transference of energy by means of a wave, which can even transverse a vacuum.

☞ A 25.

Exposure to light turns the crystals into the visual display of a static or mobile happening—better known as a photograph.

☞ A 26.

You and I and all matter. Leptons (deriving from the Greek for “weak”) and hadrons (from the Greek for “strong” or “heavy”) are two basic kinds of elementary particles. Leptons are particles that are found outside the nucleus while hadrons, like neutrons and protons, exist inside the nucleus. Orbiting electrons distancing themselves from the nucleus are leptons. Mu and tau mesons are also particles, like electrons, but they are heavier. Why electrons and their heavier twin particles, mu and tau, would want to repeat themselves in various degrees, remains a puzzle to physicists that has not yet been solved.

☞ A 27.

1) Otto Hahn (1879–1968), the German physical chemist who discovered nuclear fission and won a Nobel for chemistry (for splitting the uranium atom) in 1944.

2) Carl Friedrich von Weizsäcker (b. 1912, brother of the first post-war president of a united Germany, Richard von Weizsäcker, b. 1920), atom scientist, astrophysicist, and philosopher.

☞ A 28.

Because when living things die they cease to absorb any more carbon from the carbon dioxide in the air around them. The radioactive decay of one carbon isotope (carbon-14) from what used to be alive is constant at a rate of half every 5,700 years. This fact is the basis of the scientific technique known as carbon dating, which can date to within a few years plants and animals that lived tens of thousands of years ago.

☞ A 29.

It is one particular crystalline form of the element carbon. A) Carbon occurs as charcoal, which is the main component of coal, which keeps us warm. Charcoal is wood that has been partially burned under turf for millions of years. B) What we share with automobiles is the fact that, combined with oxygen, mankind and automobiles both "exhale" carbon dioxide. C) In rare cases crystalline carbon appears as a diamond. It can be made to combine with oxygen. In fact, if a diamond is burned up in an electric arc it will vanish in a puff of smoke as carbon dioxide.

☞ A 30.

The two engineers are Sir Frank Whittle (b. 1907) and Germany's Hans J.P. von Ohain (b. 1912). They are the fathers of the jet engine, which they invented and developed independently about sixty years ago. Whittle invented the jet engine in 1930 and ran it successfully in a lab in 1937, but von Ohain was the first in the air with a rather primitive jet-powered plane when his test pilot flew a Heinkel 178 in August 1939, just before the outbreak of the war. The British made their first test flight in 1941. For their invention, both scientists were awarded the 1991 Charles Stark Draper Prize (an award regarded as the Nobel of engineering), sharing a purse of $375,000.

☞ A 31.

Both were pioneers of photography. Scott-Archer, a sculptor, invented the collodion or wet-plate process in 1851. In 1871, Maddox, a medical doctor, invented halide gelatin emulsion, which brought about the manufacture of dry plates and decreased exposure times to less than a second. These two men were preceded in the 1830s by the more famous Louis Jacques Mande Daguerre who developed the daguerreotype, the first permanently fixed photographic image.

☞ A 32.

The scientist is Germany's Georg Simon Ohm (1787–1854). The SI (Système International) unit, one of seven scientific units used by scientists worldwide, such as the meter (m) for length, kilogram (kg) for weight, ampere (A) for electrical current, kelvin (K) for temperature, etc., of electrical resistance is named after him, but so is the unit of conductance—the reverse of resistance—which is known as mho. Ohm's law (1827) states that the steady electrical current in a metallic circuit is directly proportional to the total voltage and inversely proportional to the resistance. More simply put: the current of electricity is equal to the ratio of the voltage to the resistance.

☞ A 33.

Not at all. Instead of radiation sickness, you've got an acronym. Read the first letter of every noun and adjective in this question from "light" to "radiation" and you've got the word *laser.* Laser beams really are light waves that have been brought together in step with one another to produce a very narrow powerful monochromatic beam of great intensity. Using laser beams, surgeons, for instance, can cut living tissue and perform minute operations with a high degree of accuracy.

☞ A 34.

Albert Einstein (1879–1955) to his Serbian coworker Mileva Marić, in 1901. *Our* work referred to the relativity theory. The lovers (their illegitimate daughter was given away for adoption) shared classes at the Swiss Polytechnic (Swiss Federal Institute of Technology) in Zurich, later married, had two children, and separated in 1914. (Einstein paid her his Nobel money as part of his alimony.) Their love letters were published in 1987. But Mileva, who died in 1948, never published a scientific paper alone under her own name. It remains, however, a puzzle why Einstein never explained where he got the idea for relativity. In 1990, *Time* magazine claimed that a Russian physicist, Abram Joffe, reported in a Yugoslav biography of Marić that the original 1905 papers he had seen were signed Einstein-Marić, but that, evidently, these were the only scientific papers she had ever signed.

☞ A 35.

It is known as chloroform. The Greek word for green is *chloros,* and the Latin word for ant is *formica.*

☞ A 36.

This tiny part of a second is the exact amount of time that light travels one meter through a vacuum. This is important because in 1984 the General Conference on Weights and Measures in Paris decided to base the standard length of a meter on the speed of light. This definition is expected to remain unchanged for a considerable amount of time, according to the National Institute of Standards and Technology.

☞ A 37.

The name of this 1714 work is *Monadology.* Its writer was the German mathematician and philosopher Gottfried Wilhelm Leibniz (1614–1716). Like virtually all his works, he wrote it in Latin, including the calculus, which he developed independently of and concurrently with Newton (1642–1727). It is an irony that Leibniz, a great philosopher and mathematician, died impoverished and neglected, with only his secretary attending his funeral in Hanover, while Sir Isaac Newton was buried in Westminster Abbey in pomp and splendor the day after he died.

☞ A 38.

Dry ice is not like regular ice, which melts in warm temperatures. The solid state of dry ice converts immediately to vapor and never experiences a state that can be considered a liquid condition.

☞ A 39.

The English names derived from the Greek words are *hyperbola* (exceed), *parabola* (equal), and *ellipse* (fall short). Apollonius (c. 261–190 B.C.) also worked out the mathematical formulas for describing these conic cross-sections. Apollonius was honest enough to admit that he made use of the four books on conics by Euclid (c. 330–260 B.C.) and of other writers' works as well.

☞ A 40.

Much earlier. It was mentioned by Thomas of Cantimpré, also known as Thomas Brabantinus (Brabançon), in his encyclopedic nineteen-volume work *On the Nature of Things* (c. 1228–1244). Cantimpré was born between 1186 and 1210 and he died between 1276 and 1294. Today he is best-known as a natural historian, encyclopedist, and theologian.

☞ A 41.

When it was discovered in the 1940s that Turing was a homosexual he was given two choices: two years in prison or an experimental course of hormone treatment with estrogen injections to suppress his libido. He chose the latter, which caused him to grow breasts and atrophied his genitalia. Two years later, on June 7, 1954, he died by eating a cyanide-coated apple. One of his favorite films had been Disney's *Snow White and the Seven Dwarfs.* He was forty-one years old.

☞ A 42.

Fans hardly cool the air, if at all. In fact, the frictional heat generated by the rotating blades may even cause the temperature to rise slightly. However, the air flow that the fan draws through the blades produces a faint pressure difference in the air from one side of the blade to another. The primary result is that the blades mix the hot air at the top with the cooler air at the bottom. It is this stir in the air that cools the skin by drying and eliminating the perspiration on a person's body. Thereby body temperature is lowered a fraction.

☞ A 43.

"Brix" is the winemakers' measure of sugar in grapes. Between July and September and October, the sun ripens the grapes, which develop additional sugar as a result of photosynthesis. The measurement is named after A. F. W. Brix, a nineteenth-century German inventor.

☞ A 44.

The flight was four miles an hour faster. The average speed of *Voyager* was 110 miles per hour. Lindbergh's was just over 106 miles per hour.

☞ A 45.

Nobody knows what Einstein's last words were. The hospital had failed to assign a German-speaking nurse to him. And since the great physicist's final words were uttered in German, they were not understood.

☞ A 46.

He invented the aqualung.

☞ A 47.

They failed to realize that the thorium was changing into an isotope of radium. Indeed, it was the 1921 Nobel winner for chemistry, Frederick Soddy, who coined the term *isotope* in 1913. It described one of a species of atoms of one chemical element which have the same atomic number and are nearly identical in chemical behavior but vary in atomic mass and weight, meaning that they differ in the number of neutrons they contain.

☞ A 48.

False. *Philosophiae Naturalis Principia Mathematica* contained the theories of cosmology based on principles laid down by Newton. In fact, only one slight error has ever been found in this monumental work, an error that was not detected until three centuries after its completion in 1687. Robert Garisto, a twenty-three-year-old student at the University of Chicago, found it in the 1980s. A calculation in the book depended on the angle between two lines from the earth to the sun, but because the angle was not precisely known at the time, Newton settled on 10.5 sec., or about three-thousandths of a degree less than in his later revision. Some of his other calculations were then based on an earlier estimate of 11 sec. This inconsistency, however, has no bearing on Newton's theory.

☞ A 49.

Besides the United States, the only two countries that have not formally begun converting to the metric system are Burma and Liberia. The first proposal made by a government official to have the United States adopt the metric system was in 1790 by Secretary of State Thomas Jefferson (1743–1826). It was turned down because America's major trading source, Great Britain, had not adopted the metric system at that time.

☞ A 50.

The man was Albert Einstein (1879–1955). The person who got him interested in science was his father's brother, who brought him popular science magazines and taught him mathematics while Albert was still in elementary school. The title of his doctoral dissertation was "A New Determination of Molecular Dimensions." The four important papers he wrote dealt with the creation of the special theory of relativity, the establishment of the mass energy equivalence, the foundation of the photon theory of light, and the creation of the theory of Brownian motion. They chartered the three main directions of Einsteinian physics: the theory of relativity (meaning that gravity could be explained as "curved" space-time), statistical mechanics, and the quantum theory of radiation.

☞ A 51.

The famous figure who accused Lavoisier (1743–1794) was none other than Marat, who claimed in 1792 that Lavoisier not only had a wall put around Paris, but that this wall deprived its residents of air. Lavoisier was arrested and tried on May 8, 1794. He was guillotined in the afternoon on what is now the Place de la Concorde. Marat could not witness the execution. He had been murdered the year before by Charlotte Corday in his bath, which he had taken to ease the pain of his skin affliction. Four days later Charlotte Corday was guillotined. To add insult to injury, the Nazis destroyed a Parisian statue of Lavoisier during their occupation of Paris in World War II.

☞ A 52.

The quadrangle, or quadrilateral, is a four-angled shape, derived from the Latin *quattuor.* The Latin for three is *tres*—the triangle. The Greek-derived English equivalents for these two figures—tetragon and trigon—are virtually never used.

☞ A 53.

What these four seemingly unrelated items have in common is the atom bomb.

1) A curie is the unit of radioactivity.

2) *Enola Gay* was the name of the plane that dropped the radioactive atom bomb on Hiroshima.

3) The Italian at the U.S. Navy Department was the physicist who proved the existence of new radioactive elements, Enrico Fermi (1901–1954).

4) The scientist was Ernest Rutherford. Born in New Zealand, he became one of the world's leading physicists and won the 1908 Nobel Prize in chemistry. He is best known for the theory that the atom is not indivisible and consists of a nucleus surrounded by electrons revolving in planetary orbits.

☞ A 54.

The scientific contraption was the telegraph. The inventor was Samuel Finley Breese Morse (1791–1872), who for most of his life had been an artist of historical paintings. Even though electric telegraphs had already been proposed before 1800, Morse's, completed in late 1835, was the first working model. In 1836, he had developed the Morse code (with Alexander Bain), which was slightly altered later and is still being used today. Although rival inventors sued Morse for the rewards of the telegraph, the United States Congress voted in 1843 to pay him to build the first telegraph line in the United States from Baltimore to Washington. In 1854, the Supreme Court confirmed that the patent rights for his invention were valid and his alone.

☞ A 55.

Not always. Sometimes it does, at other times it doesn't. Boiling water rises and expands, running over the rim of the pot. Nevertheless, as temperatures get colder, water pipes outside the house have a tendency to burst because the water freezing in them also expands. But when you heat water from a temperature well below freezing (0° to 4° C) it does not expand but actually draws together, which means that it is denser at 4° than at 0°. That explains why at its densest, water well below the ocean's surface can be warmer than near the surface, which enables some fish to survive at these lower depths.

☞ A 56.

The Latin title, under which it was first published in 1687, is *Philosophiae Naturalis Principia Mathematica,* or *Mathematical Principles of Natural Philosophy,* generally shortened to *The Principia.* Its author is Sir Isaac Newton (1642–1727). He wrote the three-volume book (with the aid of Edmund Halley) in eighteen months and coincidentally laid the foundation of modern science.

☞ A 57.

It is not the leaves or branches that attract lightning but whatever is the tallest thing on a reasonably level ground. Just a single tree in the middle of a meadow will conduct electricity. Another reason is that moisture serves as a good conductor of electricity in natural surroundings, specifically the sap underneath the bark of a tree.

☞ A 58.

No. He concentrated too much on the anatomy of birds' wings, which was only effective for mere gliding purposes. It did not occur to him that to be propelled forward, some internal combustion engine had to be applied to a jet or a propeller that beat the air not only downward, but backward. His sketches of a pseudo-helicopter mechanism, which was a swiftly turning screw that would spiral its way in the air, came closer to the secrets of modern flight.

☞ A 59.

This engine was used for the first time for pumping water from mines.

☞ A 60.

The scientist was Lise Meitner (1878–1968). She was an Austrian physicist of the Jewish faith who was protected by physics-chemist Otto Hahn. They worked together from 1907–1938. She had detected radiothorium and protactinium in 1917, but Hitler forced her out of Nazi Germany after she had served his purpose. Meitner and the Jewish scientist Otto Frisch interpreted the bombardment of uranium and its consequent formation of barium as the splitting of the uranium nucleus into two almost equal parts. In 1938 her detection of thorium and uranium fission proved to be the basis for all methods to tap atomic energy, which included the development of the atomic bomb. This information was conveyed to the Danish Nobel physicist Niels Bohr (1885–1962) after Meitner's arrival in Copenhagen in 1939. Bohr left for the United States, conveying the discovery to Einstein, and Lise Meitner fled to freedom in Sweden, where she lived to be ninety.

☞ A 61.

Plato (c. 427–347 B.C.).

☞ A 62.

Carbon (C). About 800,000 carbon compounds are known to exist in chemical literature, and carbon is found in both combined and free states.

☞ A 63.

It led to Rutherford's correct observation in 1911 that the positively charged atomic nucleus was surrounded by electrons bearing a negative electric charge. His concept was confirmed two years later in experiments conducted by Niels Bohr, Hans Geiger, and Ernest Marsden.

☞ A 64.

The perfect height is 22,300 miles above the equator.

☞ A 65.

Almost exactly a century earlier, in 1674, John Mayow, an English lawyer and physiologist (1640–1679), discovered oxygen in all but name. When he placed a mouse and a candle on a raised platform in a bowl of water and topped all of this with a glass bell, he could watch the water rise as the atmospheric pressure fell inside the bell. After the water level rose about an inch, the mouse expired and the candle went out. Although there was plenty of air in the vessel, the component that sustained life in the mouse and the flame was burned up. He called that substance "nitro aerial spirit." Joseph Priestley called it dephlogisticated air, and Lavoisier gave it the name that stuck: oxygen.

☞ A 66.

About 1741, England's Royal Society and the French Academy of Sciences decided to collaborate to produce measurements and weights in terms intelligible to everybody. Two brass rods in conformity with the English standard yard kept in the Tower of London were sent to Paris. One was kept by the French, the other was returned to England engraved with a graduated half-toise (three Paris feet). England's celebrated scientific instrument maker, George Graham (1673–1751), had Jonathan Sisson (1690–1747) divide both the yard and the half-toise once more, this time into thirds, or feet. A more precise "No. 1 standard yard" then was cast on gold studs and sunk into a specific bronze bar in 1845.

☞ A 67.

When he became a citizen of the United States in 1894 he changed his name from Karl Rudolf Steinmetz (1865–1923) to Charles Proteus Steinmetz. Among his many inventions and 195 patents are the induction regulator and the magnetic arc-lamp electrode, which brought major improvements in the construction of motors, generators, and transformers, and the modernization of high-voltage, alternating-current transmission techniques. He was an early advocate of atmospheric pollution control, and he did research on solar energy conversion, the electrification of railways, and the synthetic production of protein.

☞ A 68.

Quattuordecillion is a cardinal number represented in the United States and France by a one followed by forty-five zeros. In Great Britain and Germany the one is followed by eighty-four zeros.

☞ A 69.

There is so much salt in the Dead Sea because that body of water has no outlet. The only salt that does vanish from the Dead Sea is the small amount that evaporates. While the salt found in the Atlantic is 31 pounds per ton of water, the Dead Sea contains about 187 pounds of salt per ton of water. With such a high percentage of the water consisting of salt, the Dead Sea cannot support fish life.

☞ A 70.

Goethe (1749–1832) argued that students of nature must not transform what they see into concepts and then into words. They must think only in terms of what they see. This theory countermands today's scientific concepts. Moreover, Goethe disliked mathematical physics. He insisted, erroneously, that concrete phenomena can be expressed in mathematical formula only if their essential conditions are not known. From this premise derive abstractions, which do not tend to lead to an absolute in logic (which he preferred to articulate in words), and which consequently produced flawed conclusions.

☞ A 71.

The scientist was Max Planck (1858–1947). He is most honored today for his quantum theory.

☞ A 72.

The number –273.15 (–459.69° F) represents absolute zero on the Centigrade scale. Hypothetically, this is the point at which a substance would have no molecular motion and no heat.

☞ A 73.

The physicist was Heinrich Hertz (1857–1894). When Hertz enlarged the first coil (the transmitter), placing a mobile sheet of metal behind it, he deflected electromagnetic and light waves from their path and redirected them to other coils (receivers). He became the first person to provide practical proof of the electromagnetic wave theory of light of Great Britain's Faraday and Maxwell. Shortly before Hertz's death, at thirty-seven, he discovered the progressive propagation of electromagnetic action through space, and only two years later Marconi and Popov transmitted radio signals over distances of several miles. Without realizing it, Hertz had discovered the waves that relay radio transmissions. A unit of frequency equal to 1,000 cycles per second is called kilohertz (kHz).

☞ A 74.

In the kitchen. The end result of this chemical process is Teflon and other nonstick coatings.

☞ A 75.

If the boy was born on February 29, 1896, his first birthday would be February 29, 1904. The leap year did not take place in 1900 because centuries only divisible by 400 are leap years and the boy could consequently not celebrate his first birthday until February 29, 1904, and his second until 1908 when he became twelve years old.

☞ A 76.

It is neither. The formula is a chemical used for about 95 percent of general dry cleaning. It evaporates quickly and is used in closed cleaning machines where soiled articles are revolved in the solvent until stains are removed. Dry cleaning really is a misnomer since it is only dry in the sense that the various chemicals used complete their cleaning mission without water while the main chemical ingredient is a solvent, a liquid that dissolves a substance.

☞ A 77.

The scientist was the Italian mathematician Niccolò Tartaglia (1499–1559). He found the theoretical answer that the angle of elevation at which a cannon would achieve its greatest range was forty-five degrees. A test confirmed this mathematical prediction. However, Tartaglia had his work *Nova Scientia* finally published in the firm belief that it would save his own countrymen when Venice, in 1537, was convinced that there would be an imminent invasion by Turkish Muslims. Tartaglia's book can be considered the first scientific text on ballistics and projectile motion.

☞ A 78.

The art of photography was born. Louis Daguerre (1789–1851) had invented the process of photography that carries his name—daguerreotype. In all fairness it must be stated that his coworker, the little-known J. Nicéphore Niepce, had worked with Daguerre on their "heliographic pictures" from 1829 until the former's death in 1833. This prompted the French government to institute a law in 1839 that assigned to Daguerre and the heir of Niepce annuities of 6,000 francs and 4,000 francs, respectively, provided that the theory of their photographic process be transmitted to and approved by the French Academy of Sciences, which recognized it as a solid achievement.

☞ A 79.

The electrons seek an escape route, and the most common method to reach the surface of the Earth is by a lightning bolt. Good lightning conductors include tall trees, TV antennas, radio towers made of steel, etc., or anything, regardless of its composition, that is taller than its surroundings.

☞ A 80.

It is indeed. In fact, that is the gist of Werner Heisenberg's formulation of his quantum mechanics, the theory that energy transferences occur in bursts of a minimum quantity. Simply put, Heisenberg's theory proves that atoms and elementary particles do not behave according to Newtonian mechanics.

☞ A 81.

The theoretical physicist was Richard Feynman (1918–1988). The subject he claimed nobody really understood and for which he won the Nobel Prize for physics in 1965 was the theory of quantum electrodynamics (two other physicists working independently, Julian Schwinger and Shin'ichero Tomonaga, shared the prize that year). The theory explained the almost indecipherable behavior of atoms, light, magnetism, subatomic particles, and electricity. Although he joined the faculties of Cornell University and the California Institute of Technology, Feynman rarely taught classes but simply lectured to his students; however so brilliantly that his three volumes of *Lectures on Physics* are among the most honored and magnificent physics texts of the century. And it was Feynman who demonstrated, shortly after the *Challenger* tragedy, how the shuttle's O-ring seals were responsible for the disaster. He simply dipped the booster rocket's seals into ice water, then squeezed them to prove that they lacked resilience and were bound to crack in the icy atmosphere thousands of feet above ground.

☞ A 82.

False. Electrons should not be mixed up with the speed of light. Electrons, always colliding, in this case move sluggishly. First they move in one direction in the wire, then in reverse, alternating back and forth. Traveling less than one inch per second, this alternating current (AC) may never get far from the light switch.

☞ A 83.

Because of the different wavelengths that constitute the white of the fluorescent light in a store and the brightness of sunlight on the street. The difference between these two mixtures becomes clearly visible when light is bounced back by your new suit in the open after having just been reflected to your retina under the artificial fluorescent tube.

☞ A 84.

The nature of this echo is the faint electromagnetic radiation that most likely permeates the entire universe (although it will become increasingly diluted) at a cosmic background temperature of 3°K (3 degrees above absolute zero). It stems from the Big Bang and is referred to as the cosmic microwave background. This radiation is so uniform that there is virtually no variation at present in the universe down to an accuracy of 0.1 percent. However, cosmology has not come up with an exact reason for the existence of this persistently uniform static hiss of radiation throughout the ages. Nevertheless, two Nobel winners for physics in 1978, Arno A. Penzias and Robert W. Wilson, in their 1965 thesis on electromagnetic radiation, and their 1992 discovery of cosmic "ripples" in this radiation, came close to explaining this mystery. Their accomplishment, corroborated by measurements of cosmic radiation at several wavelengths since then, has proven the energy-wavelength distribution of a cooling, ever-expanding universe, and it is one of the most highly celebrated revelations of this century's scientific disclosures in physics.

☞ A 85.

The professionals are photographers. When they refer to 18 million pixels, or picture elements, they may speak rather disparagingly of the new Photo Compact Discs whose images can be viewed on a TV screen using a special CD player. The 20 million silver molecules, on the other hand, form the main element of the standard 35mm negative while the resolution determined by 100 million silver molecules can be seen in high-quality Kodachrome slides.

☞ A 86.

Ada Byron (1815–1852), the daughter of the English poet Lord Byron, and later known as Countess of Lovelace. In her late twenties, her forty-page "Notes" was published in *Taylor's Scientific Memoirs* (1843), telling the early Victorian world about a revolutionary machine within man's grasp. She called it the Analytical Engine. Her thesis was discovered in 1953 by B. V. Bowden in Manchester, England, and her working instructions and programming analysis were so precise and correct that in 1979, a century and a quarter after her death, the United States Department of Defense named its new standardized computer language ADA in her honor. After giving birth to three children, Ada Byron died of uterine cancer at the age of thirty-seven.

☞ A 87.

The scientist who wrote this thesis was not Einstein (1879–1955) but Galileo (1564–1642) in *Dialogue Con-*

cerning the Two Chief World Systems. He wrote it in defense of Copernicus (1473–1543) about four hundred years ago.

☞ A 88.

If you study a family of elementary particles, the upsilon mesons of the question on page 74, you'll find that these mesons are bound states of a heavy quark and antiquark (particles thought to be the fundamental units of other subatomic particles). It is the binding force forming the upsilons that carries a new quantum number known as *beauty* or *bottomness.* Anti-b-quarks consequently carry antibeauty, erasing beauty, and the upsilons as such carry no beauty and are referred to as hidden-beauty states.

☞ A 89.

It is the name Einstein initially preferred to call his new Theory of Relativity. He sincerely believed that the term *invariants* presented a more precise description of his thoughts than the word *relativity.* While the phenomena of an experiment may be "relative," the laws governing physics are not; they are always constant regardless of how an observer witnesses the experiment, even though the final scientific deductions are governed by the same fixed "invariant" laws. You may not know that Einstein did not receive the Nobel Prize for his theory of relativity. The 1921 Nobel Prize was awarded to him for his discovery of the law of the photoelectric effect.

☞ A 90.

The culprit is called methane (CH_4). Although its lifetime in the atmosphere is only a decade or so, scientists believe that the deliberate setting of forest fires and the continued burning of fossil fuel, fields, and garbage will change our climate for the worse and increase the deteriorating greenhouse health hazards. However, chlorofluorocarbons and carbon dioxide account for the remaining 75–80 percent of global warming.

☞ A 91.

No, not Einstein (1879–1955), but the German mathematician Hermann Minkowski (1864–1909). In fact, it was Minkowski who simplified the mathematical description and the content of the Special Theory of Relativity by interpreting it geometrically. This provided the clues that enabled Einstein to generalize his theory, including the part dealing with gravitational forces. Minkowski's mathematical interpretation actually supplied the correct mathematical means: tensor calculus and differential geometry, that in a mathematical sense made Einstein's theory irrefutable. Even today his mathematical theory is still known among scientists as Minkowski's World or Minkowski's Universe: namely that the fourth coordinate in four-dimensional space is time and that a single event in that space is represented as a point. By the way, his brother Oskar (1858–1931) was the well-known German physiopathologist who discovered (in 1889) that a deficient endocrine secretion of the pancreas will result in diabetes.

☞ A 92.

Oil. It is assumed that oil was formed from the bodies of animals and plants millions of years ago. Dying by the billions, they sank to the bottom of the waters covering most of our planet. As the waters receded over tens of millions of years, the heat and pressure resulting from eons of weight by overlying rock, plus the fact that organic life gradually decomposed, are believed to have given birth to underground oceans of oil. That is why oil or petroleum as well as coal and natural gas are called fossil fuels.

☞ A 93.

Iron.

☞ A 94.

These are among the many mind-crushing challenges that have bedeviled mathematicians for decades and are still in search of a solution. Fermat's Last Theorem was one of them for more than 350 years, until England's Andrew Wiles, at the age of forty, almost unraveled mathematic's greatest unsolved mystery in a Cambridge (England) lecture in June 1993. The theorem actually states that there are no positive whole numbers that can solve the equation $x^n + y^n = z^n$ when n is greater than 2.

☞ A 95.

The two figures stated in the question actually constitute the same temperature, and they are important because those degrees tell the absolute zero point of temperature at which all molecular motion ceases. It is also referred to as zero degrees Kelvin, which is defined on the Celsius scale (having replaced the centigrade scale for scientific purposes). Incidentally, the boiling point on the Celsius scale was never quite 100° C, but almost exactly 99.97° C.

☞ A 96.

Baking soda, baking powder, or yeast will cause the batter to rise faster at 1,000 feet above sea level than it does below 500 feet. The reason is that the atmospheric pressure, i.e., the blanket of air, is lighter at the higher altitude. There is less resistance from the surrounding air for the carbon dioxide, and in consequence the cake can rise with more force and greater rapidity than at a lower altitude. Bakers in higher altitudes use less of these items in order to prevent cakes from ending up overrisen and therefore tasteless and tough. In fact, at even higher altitudes of 3,500 feet or more there are special commercial mixes that direct bakers to add more flour in order to improve the taste of the cake.

☞ A 97.

All you have to do is to determine the answer as a difference of squares. In our case: 74×86 is the equivalent of $(80 - 6) \times (80 + 6)$ or $80^2 - 6^2$. The square of 80 is 6,400 and then you deduct the square of 6, which is 36. 6,400 less 36 is the result, with a total of 6,364. 74×86 equals 6,364. You can mentally repeat similar arithmetic puzzles and solutions with multiplications such as 25×35 (30^2 equals 900; 5^2 equals 25; 900 less 25 is 875, the answer for

25×35); or 62×58 ($60^2 = 3{,}600$; $2^2 = 4$; $3600 - 4 = 3{,}596$; i.e., 62×58), or 86×94 ($90^2 = 8{,}100$; $4^2 = 16$; $8100 - 16 = 8084$; $86 \times 94 = 8084$), ad infinitum, and astonish your friends.

☞ A 98.

Werner Carl Heisenberg (1901–1976). The 1932 Nobel Prize–winner in physics was an originator of quantum mechanics and the formulator of the Heisenberg Uncertainty Principle.

☞ A 99.

The first puzzle is solved by putting the nine barrels in an **L**-formation, with five being placed north to south, and the remaining four placed either to the left or to the right of the bottom (most southerly) one. For the second puzzle, you form nine barrels into an **L**-shaped right angle, with five on each side, and then you place the tenth and eleventh barrels on top of the one where the perpendicular and horizontal lines of the **L** meet. Both the horizontal and perpendicular lines will now contain seven barrels per row.

☞ A 100.

Those more familiar with the latter part of the question will immediately jump to the conclusion that all three parts of the question are associated with the work of Omar Khayyám (c. 1050–1123), the Persian astronomer and poet. As a matter of fact, the best-known line of Omar Khayyám's *The Rubaiyat* was not written by the Persian but is part of an extremely free English translation (1859) by Edward FitzGerald (1809–1883), which is considered more a creation of the translator than of the original writer. However, the Persian also happened to be a mathematician who was responsible for a step toward the unification of algebra and geometry that finally came to fruition with the work of René Descartes (1596–1650) and Pierre de Fermat (1601–1665). Even though Khayyám's cubic equation of the third degree was completely geometrical and resulted only in positive roots since there are no line segments with a negative length, the sixteenth-century mathematician Tartaglia is sometimes given credit for this feat, even though his work was published in Girolamo Cardano's (better known in English as Jerome Cardan) *Ars magna* 450 years later, in 1545. Omar also assisted in reforming the Muslim calendar.

☞ A 101.

Leonardo da Vinci designed the first helicopter. However, it was never built, and if it had been built it probably would not have worked. Wan Hu, on the other hand, actually built what he designed, the first flying machine. He attached forty-seven gunpowder rockets to the back of a chair and served as his own pilot, but when the rockets were lit they exploded and killed him.

☞ A 102.

All these scientific achievements took place in 1784.

☞ A 103.

This feat was accomplished by the German physicist Rudolf Mössbauer (b. 1929) between 1958 and 1960. He discovered that it was possible to produce gamma rays with an extremely narrow wavelength by stimulating a nucleus to emit sharply defined beams of these gamma rays. This enabled him to measure correctly any change in wavelength. His theory became known as the Mössbauer effect and it was used immediately (1960) to provide the first lab test to confirm Einstein's General Theory of Relativity. For his work on gamma rays Mössbauer and the U.S. physicist Robert Hofstadter (b. 1915) won the Nobel in physics in 1961.

☞ A 104.

Stoichiometry is the term used for the branch of chemistry that deals with the atomic-weight proportions of combining elements. Some scientists call it the mathematics of chemistry. Not to be mixed up with chemical formulas, stoichiometry analyzes chemical measurements such as the structure of molecules, of the elements' atomic sizes and weights (even of vapor and gas volumes), and the weight relations in any chemical reactions. In different branches of science stoichiometry has slightly different analytical meanings. In analytical chemistry it is concerned with weight relationships in gravimetric (weight) analysis. In volumetric analysis, the volume relationships are the determining factors of stoichiometry.

☞ A 105.

Yes, the United States Navy honored him about a century after he wrote *Twenty Thousand Leagues Under the Sea* (1870). The reason: The navy credited the author in part for the successful development of its nuclear submarines.

☞ A 106.

To start with, the merchant fills up the ten-gallon bottle with wine. Next he empties six gallons from it into the six-gallon bottle, filling it completely and leaving four gallons in the ten-gallon bottle. For his third step he empties the entire six-gallon bottle into a large, unmarked container, leaving the six-gallon bottle empty. Next he empties the last four gallons from the ten-gallon bottle, leaving the ten-gallon bottle empty this time. However, he fills up the ten-gallon bottle at once to the rim and for his next step also fills the six-gallon bottle, which already contains four gallons of wine, by emptying two gallons from the ten-gallon bottle into it, leaving the ten-gallon bottle now containing the eight gallons of wine the customer wants to buy.

☞ A 107.

Simply put: nowhere. He never entered college or passed an exam to obtain a science degree, or any other degree for that matter. After three months in a Port Huron, Michigan, public school, Edison left school for good at the age of twelve, in order to become a railroad newsboy. Despite the early end to his formal education, Edison took out more than a thousand patents in his lifetime, and in 1927 was admitted to the National Academy of Sciences.

☞ A 108.

The rooster's head will point in the direction the wind is blowing to. The tradition of using the representation of a rooster on weathervanes stems from Christ's prediction that Peter would betray Him three times before the cock crowed. Henceforth the rooster on church spires became a symbol equating the fickleness of Peter's faith with the unpredictable nature of the wind.

☞ A 109.

They were referring to the shape of snowflakes. Yet none of the scientists could explain correctly why snowflake types like "plates" and "stellars" essentially have six sides. Snow is the solid form of water that grows while floating in the free air of the atmosphere. Icy atmospheric conditions convert the water into hexagonal lattice, meaning that the condensation of water on its surface becomes a symmetrical hexagonal crystal. It was only the explanation in 1784 by Henry Cavendish (1731–1810) for the chemical composition of water that helped to solve the snowflakes' symmetrical puzzle: the angle extended under the two hydrogen atoms in the middle of the oxygen atom is about 120°. The hexagonal shapes of each flake differ because the flakes' corners attract new water molecules. Also, while the flakes fall through a multitude of atmospheric regions, the meteorological conditions that the flakes encounter provide them with different hexagonal shapes, although their symmetrical forms will be preserved by the growing crystal whose external ends are subject to identical atmospheric manipulations.

☞ A 110.

The writers were Bertrand Russell (1872–1970) and Alfred North Whitehead (1861–1947). Their colossal three-volume work *Principia Mathematica* was completed after three years in 1913. Although the work's attempts to derive all mathematics from pure logic did not obtain general acceptance, Russell did find an ingenious way of overcoming some problems in the particular case of mathematical induction in the second edition (1925) of *Principia Mathematica*. The thesis of this work was only gradually accepted, but it has largely transformed the conception of logic.

☞ A 111.

It's the magnetic compass used for navigation. The first mention of a European mariner's compass did not come until a century later, when Alexander Neckam wrote about one in his *De utensilibus* (1187).

☞ A 112.

It was Euclid (c. 330–260 B.C.). Nine of his books deal with plane and solid geometry, four with number theory. Euclid, a Greek mathematician who lived in Alexandria, called them *Stoicheia/Elements*. His philosophy is based on a limited number of axioms and it is from them that many propositions are derived by the use of his logical rules. He also expounded the theory of incommensurables (having no common measure with another integral or fractional number or quantity). By the same token it is known that Euclid incorporated several discoveries of Eudoxus and Theaetetus into his own books V, X, XII and XIII, while book VII, analyzing the foundation of arithmetic, may have been partly written a hundred years earlier.

☞ A 113.

First he lays the piece of trunk on its side (the circular, flat part) and cuts it in two in the middle; then he sets the trunk again up straight, one half on top of the other half, and cuts the two flat pieces into four equal parts, using his saw only twice for the latter assignment.

☞ A 114.

There have been many claims either way throughout the decades. The true verdict arrived in 1991, when Dr. Robert D. Ballard, a scientist at Woods Hole Oceanographic Institute on Cape Cod, used a super-low light underwater electronic still camera that is more powerful than floodlights to locate the Nazi battleship almost three miles under the Atlantic. Not only could one distinctly see the *Bismarck*'s swastikas, some guns, and boots, but most important it was clear that Hitler's favorite battleship had not been crushed. Crushing happens only when the unfilled spaces inside a sinking vessel are imploded by deep-ocean pressures. This discovery supported the claims of crew members who were saved by the British that Lütjens had time to fully flood the damaged ship so it could sink, with all hatches sealed, before the British could serve it its coup de grâce. Incidentally, the Kristof-Ballard mercury-vapor lamps, which have already explored the *Titanic*, several battleships sunk at Guadalcanal, and a schooner that went down in a squall on Lake Ontario in 1912, will soon be superseded by even more powerful laser deep-sea cameras.

INDEX

*Page numbers in **boldface** refer to answers.*

INDEX

INDEX

INDEX

INDEX

INDEX

INDEX

INDEX

INDEX

INDEX

INDEX

INDEX

INDEX

INDEX